STUDENT SOLUTIONS MANUAL FOR

Elements of Physical Chemistry

With Applications in Biology

Third Edition

Charles A. Trapp

W. H. Freeman and Company
New York

ISBN: 0-7167-3897-X

Printed in the United States of America

Third printing, 2003

Contents

Introduction

Treat all gases as perfect unless instructed to do otherwise.

1. Refer to Table 0.1 for pressure conversion factors.

(a) $p = 110\ \text{kPa} \times \dfrac{760\ \text{torr}}{101.325\ \text{kPa}} = \underline{824\ \text{torr}}$

(b) $p = 0.997\ \text{bar} \times \dfrac{100\ \text{kPa}}{1\ \text{bar}} \times \dfrac{1\ \text{atm}}{101.325\ \text{kPa}} = \underline{0.984\ \text{atm}}$

(c) $p = 2.15 \times 104\ \text{Pa} \times \dfrac{1\ \text{kPa}}{10^3\ \text{Pa}} \times \dfrac{1\ \text{atm}}{101.325\ \text{kPa}} = \underline{0.212\ \text{atm}}$

(d) $p = 723\ \text{torr} \times \dfrac{101.325\ \text{kPa}}{760\ \text{torr}} \times \dfrac{10^3\ \text{Pa}}{1\ \text{kPa}} = \underline{9.64 \times 10^4\ \text{Pa}}$

3. The thinness of the Martian atmosphere is a result of the low gravitational attraction, which makes it impossible for Mars to maintain an atmosphere. The two factors cannot be separated.

The mass of any given vertical column of gas on Mars would be the same on Earth; therefore the simple answer to this question would seem to be

$$p\ (\text{Earth}) = 0.0060\ \text{atm} \times \frac{9.81}{3.7} = \underline{1.6 \times 10^{-2}}$$

The better answer would take into account the ratio of the surface area of the two planets. The *same* atmosphere would be spread more thinly on Earth.

Question: Look up the radii of Mars and Earth and calculate the better answer referred to above.

Comment: The atmospheres of planets cannot be maintained in an equilibrium state. A complete analysis is quite complicated.

5. The external (atmospheric) pressure is greater than the internal pressure, hence

$p_{ex} - p_{in} = \rho gh$ [0.2], or

$p_{in} = p_{ex} - \rho gh = 758\ \text{Torr} - \rho gh$

ρgh can be expressed in Torr as follows:

$$\rho gh = 3.55\ \text{cm H}_2\text{O} \times \frac{1\ \text{cm Hg}}{13.59\ \text{cm H}_2\text{O}} \times \frac{10\ \text{mm Hg}}{1\ \text{cm Hg}} \times \frac{1\ \text{Torr}}{1\ \text{mm Hg}} = 2.61\ \text{Torr}$$

$p_{in} = 758\ \text{Torr} - 2.61\ \text{Torr} = \underline{755\ \text{Torr}}$

7. On the Rankine scale (see the solution to Exercise 0.6) 0 °F = 459.67 °R. Hence

$$\frac{T}{°\text{R}} = \frac{\theta}{3.7} + 459.67 = 212 + 459.67 = 671.67$$

T = 671.67 °R

9. Let haemoglobin = Hb and myoglobin = Mb

$$\text{mass of Hb} = 3 \times 10^{8}\ \text{molecules} \times \frac{4\ \text{mol Mb}}{1\ \text{mol Hg}} \times \frac{1\ \text{mol Hb}}{6.02 \times 10^{23}\ \text{molecules}} \times \frac{16.1 \times 10^{3}\ \text{g}}{1\ \text{mol Mb}}$$

$$= 3.\overline{21} \times 10^{-11}\ \text{g}$$

$$\text{Fraction Hb} = \frac{3.\overline{21} \times 10^{-11}\ \text{g}}{33 \times 10^{-11}\ \text{g}} = \underline{0.9\bar{7}} \text{ or } 97\%$$

The Properties of Gases

1

Treat all gases as perfect unless instructed to do otherwise.

1. Solve the perfect gas law $pV = nRT$ [1.1] for pressure.

$$p = \frac{nRT}{V}$$

$$n = 2.045\ \text{g} \times \frac{1\ \text{mol N}_2}{28.02\ \text{g}} = 0.07298\ \text{mol}$$

$$p = \frac{nRT}{V} = \frac{(0.07298\ \text{mol})(8.3145\ \text{kPa L K}^{-1}\ \text{mol}^{-1})(294\ \text{K})}{2.00\ \text{L}} = \underline{89.2\ \text{kPa}}$$

3. Solve the perfect gas law [1.1] for n.

$$n = \frac{pV}{RT} = \frac{\left(24.5\ \text{KPa} \times 250.0\ \text{mL} \times \frac{1\ \text{L}}{10^3\ \text{mL}}\right)}{(8.3145\ \text{kPa L K}^{-1}\ \text{mol}^{-1}) \times (292.6\ \text{K})} = 0.00252\ \text{mol} = \underline{2.52 \times 10^{-3}\ \text{mol}}$$

5. Remember that p is inversely proportional to V at constant temperature. Therefore $p_2 = \frac{V_2}{V_1} \times p_1$. Converting 1.00 L to cm^3 gives $1.00 \times 10^3\ \text{cm}^3$.

$$p_1 = 1.00\ \text{atm}$$

$$p_2 = \frac{1.00 \times 10^3\ \text{cm}^3}{100\ \text{cm}^3} \times 1.00\ \text{atm} = 10 \times 1.00\ \text{atm} = \underline{10.0\ \text{atm}}$$

7. At constant temperature [1.3] becomes $p_1V_1 = p_2V_2$

$$p_2 = \frac{p_1V_1}{V_2} = \frac{101\ \text{kPa} \times 7.20\ \text{L}}{4.21\ \text{L}} = \underline{173\ \text{kPa}}$$

9. Pressure is constant, so the same relation between temperature and volume as in the previous exercise is used. Here the volume has increased by 14%, so $V_2 = 1.14\ V_1$. $T_1 = 340$ K.

$$T_2 = \frac{1.14\ V_1}{V_1} \times 340\ \text{K} = 1.14 \times 340\ \text{K} = \underline{388\ \text{K}}$$

11. Pressure is inversely proportional to volume. There will be an increase in pressure due to the depth submerged, which can be calculated from $p = \rho gh$.

$$V_f = \frac{p_i}{p_f} V_i$$

and total pressure $p_i = 1.0$ atm

$$p_f = 1.0 \text{ atm} + \rho gh$$

$$\rho gh = (1.025 \times 10^3 \text{ kg m}^{-3}) \times (9.81 \text{ m s}^{-2}) \times 50 \text{ m} = 5.0 \times 10^5 \text{ Pa}$$

therefore

$$p_f = (1.01 \times 10^5 \text{ Pa}) + (5.0 \times 10^5 \text{ Pa}) = 6.0 \times 10^5 \text{ Pa}$$

$$V_f = \frac{1.01 \times 10^5 \text{ Pa}}{6.0 \times 10^5 \text{ Pa}} \times 3.0 \text{ m}_3 = \underline{0.50 \text{ m}^3}$$

13. (a) $V = \dfrac{n_J RT}{p_J}$

There is only one volume, so using the amount of nitrogen and p_J for nitrogen we can calculate the volume.

$$n(N_2) = \frac{0.225 \text{ g}}{28.02 \text{ g mol}^{-1}} = 8.03 \times 10^{-3} \text{ mol},$$

$$p(N_2) = 15.2 \text{ kPa}, T = 300 \text{ K}$$

$$V = \frac{(8.03 \times 10^{-3} \text{ mol}) \times (8.3145 \text{ kPa L K}^{-1} \text{ mol}^{-1}) \times (300 \text{ K})}{15.2 \text{ kPa}} = \underline{1.32 \text{ L}}$$

(b) $p = \dfrac{nRT}{V}$ where n is the sum of the amounts of each component; therefore

$$n = n(CH_4) + n(Ar) + n(Ne)$$

$$n(CH_4) = \frac{0.320 \text{ g}}{16.04 \text{ g mol}^{-1}} = 2.00 \times 10^{-2} \text{ mol}$$

$$n(Ar) = \frac{0.175 \text{ g}}{39.95 \text{ g mol}^{-1}} = 4.38 \times 10^{-3} \text{ mol}$$

$$n = (2.00 + 0.438 + 0.803) \times 10^{-2} \text{ mol} = 3.24 \times 10^{-2} \text{ mol}$$

Substituting into the equation for p

$$p = \frac{(3.24 \times 10^{-2} \text{ mol}) \times (8.3145 \text{ kPa L K}^{-1} \text{ mol}^{-1}) \times (300 \text{ K})}{1.32 \text{ L}} = \underline{61.2 \text{ kPa}}$$

15. To calculate the molar mass of the compound, we need to relate density, temperature, and pressure. Using the perfect gas law:

$pV = nRT$ and density $= \dfrac{\text{mass}}{\text{volume}}$

$V = \dfrac{nRT}{p}$ and $\dfrac{m}{\rho} = \dfrac{nRT}{p}$, $\dfrac{m}{\rho} = \dfrac{\rho RT}{p}$. Note that R must be in the same units as T and p; in this case K and kPa.

$$M = \frac{(1.23\ \text{gL}^{-1}) \times (8.3145\ \text{kPa L K}^{-1}\ \text{mol}^{-1}) \times (330\ \text{K})}{(25.5\ \text{kPa})} = \underline{132\ \text{g mol}^{-1}}$$

17. (a) The partial pressure of the gases can be related to their mole fractions. The total amount is found by $n = n(H_2) + n(N_2) = 2.0\ \text{mol} + 1.0\ \text{mol} = 3.0\ \text{mol}$

$$x(H_2) = \frac{2.0\ \text{mol}}{3.0\ \text{mol}} = 0.67$$

$$x(N_2) = \frac{1.0\ \text{mol}}{3.0\ \text{mol}} = 0.33$$

$$p_J = n_J \frac{RT}{V}$$

Solving for $\dfrac{RT}{V}$

$$\frac{RT}{V} = \frac{(8.2508 \times 10^{-2}\ \text{L atm K}^{-1}\ \text{mol}^{-1}) \times (273.15\ \text{K})}{22.4\ \text{L}} = 1.00\ \text{atm mol}^{-1}$$

$p(H_2) = 2.0\ \text{mol} \times 1.00\ \text{atm mol}^{-1} = \underline{2.0\ \text{atm}}$

$p(N_2) = 1.0\ \text{mol} \times 1.00\ \text{atm mol}^{-1} = \underline{1.0\ \text{atm}}$

(b) the total pressure is the sum of the partial pressures $p = p(H_2) + p(N_2)$

$= 2.0\ \text{atm} + 1.0\ \text{atm} = \underline{3.0\ \text{atm}}$

19. The formula for determining the mean free path is $\lambda = \dfrac{RT}{\sqrt{2}N_A\sigma p}$ [1.19], solving for p,

$$V = \frac{4}{3}\pi r^3$$

$$V = 1.0\ \text{L} = 10^{-3}\text{m}^3$$

$$\frac{4}{3}\pi r^3 = 10^{-3}\text{m}^3$$

$$r = 0.062\ \text{m}$$

$$\lambda = d = 0.124\ \text{m}$$

$$p = \frac{RT}{\sqrt{2}N_A\sigma\lambda}$$

$$p = \frac{(8.3145 \text{ Pa m}^3 \text{ K}^{-1} \text{ mol}^{-1}) \times (298.15 \text{ K})}{\sqrt{2} \times (6.02 \times 10^{23} \text{ mol}^{-1}) \times (0.36 \times 10^{-18} \text{ m}^2) \times (0.124 \text{ m})} = \underline{0.065 \text{ Pa}}$$

21. The formula for determining the mean free path is $\lambda = \frac{RT}{\sqrt{2}N_A\sigma p}$ [1.25]

$$\lambda = \frac{(8.3145 \text{ J K}^{-1} \text{ mol}^{-1}) \times (217 \text{ K})}{\sqrt{2} \times (6.02 \times 10^{23} \text{ mol}^{-1}) \times (0.43 \times 10^{-18} \text{ m}^2)} \times \frac{1}{(0.050 \text{ atm}) \times (1.013 \times 10^5 \text{ Pa atm}^{-1})}$$

$$= 973 \text{ nm} = \underline{0.97\ \mu\text{m}}$$

23. We have already calculated the number of collisions for a single Ar atom. Therefore, we need to calculate the total number of atoms to determine the total number of collisions.

The total number of molecules will be $6.02 \times 10^{23} \text{ mol}^{-1} \times n$

Because $n = \frac{pV}{RT}$,

$$10 \text{ bar} \times \frac{1 \text{ atm}}{1.013 \text{ bar}} = 9.87 \text{ atm}$$

$$\text{number of molecules} = \frac{6.02 \times 10^{23} \text{ molecules mol}^{-1} \times pV}{RT}$$

$$= \frac{(6.02 \times 10^{23} \text{ molecules mol}^{-1}) \times (9.87 \text{ atm}) \times (1.0 \text{ L})}{(0.08206 \text{ L atm K}^{-1} \text{ mol}^{-1}) \times (298 \text{ K})} = 2.43 \times 10^{23} \text{ molecules}$$

$$\text{Total number of collisions} = \frac{2.43 \times 10^{23} \text{ molecules} \times (5.3 \times 10^{10} \text{ collisions/molecules})}{2}$$

$$= \underline{6.4 \times 10^{33} \text{ collisions}}$$

The answer is divided by two because two molecules collide in one collision event.

By substituting in for the other pressures,

(b) $\underline{6.4 \times 10^{31} \text{ collisions}}$

(c) $\underline{6.4 \times 10^{21} \text{ collisions}}$

25. $\lambda = \dfrac{RT}{\sqrt{2}N_A\sigma p}$; σ is given as 0.43 nm^2

Because we want to solve this for several different p's, calculate first in terms of p so it is easier to substitute into the equation

$$\lambda = \frac{(8.3145 \text{ J K}^{-1} \text{ mol}^{-1}) \times (298.15 \text{ K})}{\sqrt{2} \times (6.02 \times 10^{23} \text{ mol}^{-1}) \times (0.43 \times 10^{-18} \text{ m}^2) \times p}$$

$$\lambda = \frac{6.8 \times 10^{-3} \text{ m Pa}}{p}$$

(a) When $p = 10$ bar $= 1.0 \times 10^6$ Pa

$$\lambda = \frac{6.8 \times 10^{-3} \text{ m Pa}}{1.0 \times 10^6 \text{ Pa}} = 6.8 \times 10^{-9} \text{ m} = \underline{6.8 \text{ nm}}$$

(b) When $p = 10^3$ kPa

$$\lambda = \frac{6.8 \times 10^{-3} \text{ m Pa}}{103 \times 10^3 \text{ Pa}} = 0.066 \times 10^{-6} \text{ m} = 6.6 \times 10^{-8} \text{ m} = \underline{68 \text{ nm}}$$

(c) When $p = 1$ Pa, $\lambda = 6.8 \times 10^{-3}$ m $= \underline{7 \text{ mm}}$

27. $\lambda = \dfrac{RT}{\sqrt{2}N_A\sigma p}$ [1.19]

At constant V, p varies directly with T

$p = \dfrac{nRT}{V}$ and

$$\lambda = \frac{RT}{\sqrt{2}N_A\sigma} \times \frac{V}{nRT} = \frac{V}{\sqrt{2}N_A\sigma n}$$, hence λ is <u>independent of temperature</u>.

29. For a perfect gas $p = \dfrac{nRT}{V}$

$$n = 10.00 \text{ g } CO_2 \times \frac{1 \text{ mol } CO_2}{44.01 \text{ g}} = 0.2272 \text{ mol}$$

$$p = \frac{(0.2272 \text{ mol}) \times (0.08206 \text{ L} \cdot \text{atm mol}^{-1} \text{ K}^{-1}) \times (298.1 \text{ K})}{0.100 \text{ L}} = \underline{55.6 \text{ atm}}$$

For a van der Waals gas:

$$p = \frac{nRT}{V - nb} - a\left(\frac{n}{V}\right)^2$$

$a = 3.59$ L^2 atm mol^{-2}

$b = 0.043$ L mol^{-1}

$$p = \frac{(0.2272\text{ mol}) \times (0.08206\text{ L atm mol}^{-1}\text{ K}^{-1})(298.1\text{ K})}{(0.100\text{ L}) - (0.2272\text{ mol} \times 0.043\text{ L mol}^{-1})} - 3.59\text{ L}^2\text{ atm mol}^{-2}$$

$$\times \left(\frac{0.2272\text{ mol}}{0.100\text{ L}}\right)^2 = 43.1\text{ atm} \times 101.325\text{ kPa/atm} = \underline{4.37\text{ M Pa}}$$

31. Because $C = 1200\text{ cm}^6\text{ mol}^{-2}$, $b = C^{1/2} = \underline{34.6\text{ cm}^3\text{ mol}^{-1}}$

$$a = RT(b - B)$$
$$= (8.206 \times 10^{-2}\text{ L atm mol}^{-1}\text{ K}^{-1}) \times (273\text{ K}) \times (34.6 + 21.7\text{ cm}^3\text{ mol}^{-1})$$
$$= (22.40\text{ L atm mol}^{-1}) \times (56.3 \times 10^{-3}\text{ L mol}^{-1})$$
$$= \underline{1.26\text{ L}^2\text{ atm mol}^{-2}}$$

Thermodynamics: The First Law

2

Treat all gases as perfect unless instructed to do otherwise.

1. To calculate the work use the formula $w = mgh$ [2.1]

   (a) $w = 1.0\ \text{kg} \times 9.81\ \text{m s}^{-2} \times 10\ \text{m} = \underline{98\ \text{J}}$

   (b) $w = 1.0\ \text{kg} \times 1.60\ \text{m s}^{-2} \times 10\ \text{m} = \underline{16\ \text{J}}$

3. $w = mgh$ [2.1]

   $= 65\ \text{kg} \times 9.81\ \text{m s}^{-2} \times 4.0\ \text{m} = \underline{2.6\ \text{kJ}}$

5. We will assume that in the complete combustion of glucose that the product CO_2 is a gas, but H_2O is a liquid; that is

   $C_6H_{12}O_6(s) \rightarrow 6\ CO_2(g) + 6\ H_2O\ (l)$

   The work of expansion is then due to the $CO_2(g)$ produced in the reaction.

   $$\text{amount } (n) \text{ of } CO_2 = 1.0\ \text{g}\ C_6H_{12}O_6 \times \frac{1\ \text{mol}}{180.2\ \text{g}} = 5.5\bar{5} \times 10^{-3}\ \text{mol}$$

   $$\Delta V = V = \frac{nRT}{p}$$

   $w = -p_{ex}\,\Delta V$ [2.2, constant pressure process]

   $p_{gas} = p_{ex}$, therefore

   $$w = -nRT = -5.5\bar{5} \times 10^{-3}\ \text{mol} \times 8.315\ \text{J K}^{-1}\ \text{mol}^{-1} \times 293\ \text{K}$$

   $= \underline{-14\ \text{J}}$

7. (a) horizontally (no additional work to raise the piston) work done by system

   = distance × opposing force

   $w = h \times p_{ex} \times A$

   $$= 155\ \text{cm} \times 105\ \text{k Pa} \times 55.0\ \text{cm}^2 \times \frac{1\ \text{m}^3}{10^6\ \text{cm}^3} \times \frac{10^3\ \text{Pa}}{1\ \text{k Pa}}$$

   $= 895\ \text{Pa m}^3 = \underline{895\ \text{J}}$

(b) vertically (additional work is required to raise the piston)

additional work = $h \times mg$ [2.1]

$$w_{\text{additional}} = 155\ \text{cm} \times \frac{1\ \text{m}}{100\ \text{cm}} \times 250\ \text{g} \times \frac{1\ \text{kg}}{10^3\ \text{g}} \times 9.81\ \text{ms}^{-2}$$

$$= 3.80\ \text{J}$$

Total work = 895 J + 3.80 J = 899 J

9. $w = -nRT \ln \frac{V_\text{f}}{V_\text{i}}$ [2.3]

$$nRT = 52.0 \times 10^{-3}\ \text{mol} \times 8.315\ \text{J K}^{-1}\ \text{mol}^{-1} \times 260\ \text{K}$$

$$= 1.124 \times 10^2\ \text{J}$$

$$w = -1.124 \times 10^2\ \text{J} \times \ln\left(\frac{100\ \text{mL}}{300\ \text{mL}}\right) = \underline{+123\ \text{J}}$$

11. $\text{Mg(s)} + 2\ \text{HCl(aq)} \rightarrow \text{H}_2\text{(g)} + \text{MgCl}_2\text{(aq)}$

$M(\text{Mg}) = 24.31\ \text{g mol}^{-1}$

$$n_{\text{H}_2} = n_{\text{Mg}} = \frac{12.5\text{g}}{24.31\ \text{g mol}^{-1}} = 0.514\ \text{mol}$$

$$w = -p_\text{ex}\,\Delta V\ [2.2]$$

$$V_\text{i} = 0,\ V_\text{f} = \frac{nRT}{p_\text{f}},\ p_\text{f} = p_\text{ex}$$

$$w = -p_\text{ex}(V_\text{f} - V_\text{i}) = -p_\text{ex} \times \frac{nRT}{p_\text{ex}} = -nRT$$

$$w = 0.514\ \text{mol} \times 8.315\ \text{J K}^{-1}\ \text{mol}^{-1} \times 293.5\ \text{K}$$

$$= -1.25 \times 10^3\ \text{J} = \underline{-1.25\ \text{kJ}}$$

13. $C = \frac{q}{\Delta T}$ [2.5] $= \frac{124\ \text{J}}{5.23\ \text{K}} = \underline{23.7\ \text{J K}^{-1}}$

15. $C_V = \frac{q_V}{\Delta T}$ [2.5] $= \frac{229\ \text{J}}{2.55\ \text{K}} = 89.8\ \text{J K}^{-1}$

The molar heat capacity at constant pressure is therefore

$$C_{V,\text{m}} = \frac{89.8\ \text{J K}^{-1}}{3.0\ \text{mol}} = \underline{30\ \text{J K}^{-1}\ \text{mol}^{-1}}$$

For a perfect gas $C_{p,\text{m}} - C_{V,\text{m}} = R$, or

$$C_{p,\text{m}} = C_{V,\text{m}} + R = (30 + 8.3)\ \text{J K}^{-1}\ \text{mol}^{-1} = \underline{38\ \text{J K}^{-1}\ \text{mol}^{-1}}$$

17. $q = I\mathcal{V}t$ [2.4] $= 1.27\ \text{A} \times 12.5\ \text{V} \times 157\ \text{s} = 2.49 \times 10^3\ \text{J}$

$1\ \text{A s} = 1\ \text{C},\ 1\ \text{C V} = 1\ \text{J}$

$$C = \frac{q}{\Delta T}\ [2.5] = \frac{2.49 \times 10^3\ \text{J}}{3.88\ \text{K}} = 642\ \text{J K}^{-1}$$

$q = C\Delta T = 642\ \text{J K}^{-1} \times 2.89\ \text{K} = \underline{1.86 \times 10^3\ \text{J}}$

19. $\Delta U = w + q$ [2.8]

$$w = mgh\ [2.1] = 200\ \text{g} \times \frac{1\ \text{kg}}{10^3\ \text{g}} \times 9.81\ \text{m s}^{-2} \times 1.55\ \text{m} = 3.04\ \text{J}$$

This is the work on the mass; the work done by the animal is –3.04 J.

$\Delta U = -3.04\ \text{J} + (-5.0\ \text{J}) = \underline{-8.0\ \text{J}}$

21. $H = U + pV$ [2.11], hence the difference between molar enthalpy and molar energy is pV.

(a) For a perfect gas $pV = nRT = RT$ (for $n = 1$)

$RT = 8.3145\ \text{J K}^{-1}\ \text{mol}^{-1} \times 298.15\ \text{K} = 2.47897 \times 10^3\ \text{J mol}^{-1}$

$= \underline{2.479\ \text{kJ mol}^{-1}}$

(b) Use the virial equation of state [1.28] with B and C written in terms of the van der Waals constants. See the solution to Exercise 1.39 for the required expressions.

$$pV_\text{m} = \frac{RT}{V_\text{m}}\left(1 + \frac{B}{V_\text{m}} + \frac{C}{V_\text{m}^2} + \ldots\right)$$

For the purposes of this exercise, we may wish to truncate this power series after the second term and approximate V_m as $\frac{RT}{p}$, then

$$pV_\text{m} = RT\left(1 + \frac{pB}{RT} + \ldots\right),\ B = \left(b - \frac{a}{RT}\right)$$

$$pV_\text{m} = RT + pb - \frac{pa}{RT} + \ldots$$

Assume $p = 1.000\ \text{atm} = 1.01325 \times 10^5\ \text{Pa}$ in the above expression.

$$pV_\text{m} = (2.4790 \times 10^3\ \text{J mol}^{-1}) + (1.01325 \times 10^5\ \text{Pa} \times 0.043\ \text{L mol}^{-1} \times \frac{10^{-3}\ \text{m}^3}{1\ \text{L}})$$

$$-\left(\frac{1.01325 \times 10^5\ \text{Pa} \times 3.59\ \text{L}^2\ \text{atm mol}^{-2} \times 1.01325 \times 10^5\ \text{Pa/1 atm} \times 10^{-6}\ \text{m}^6/\text{L}^2}{2.4790 \times 10^3\ \text{J mol}^{-1}}\right)$$

$= [(2.4790 \times 10^3) + (4.4 - 14.9)]\ \text{J mol}^{-1}$

$= 2.4685 \times 10^3\ \text{J mol}^{-1} = \underline{2.4685\ \text{kJ mol}^{-1}}$

As this value is less than the perfect gas value, the molar enthalpy is *decreased* by intermolecular forces.

23. $q = C_p\Delta T = nC_{p,\mathrm{m}}\Delta T$

$= 3.0\ \mathrm{mol} \times (29.4\ \mathrm{J\ K^{-1}\ mol^{-1}}) \times 25\ \mathrm{K} = \underline{+2.2\ \mathrm{kJ}}$

$\Delta H = q = \underline{+2.2\ \mathrm{kJ}}$ (constant pressure)

$\Delta U = \Delta H - \Delta(pV) = \Delta H - \Delta(nRT)$ (perfect gas)

$= \Delta H - nR\Delta T$

$= 2.2\ \mathrm{kJ} - 3.0\ \mathrm{mol} \times 8.315\ \mathrm{J\ K^{-1}\ mol^{-1}} \times 25\ \mathrm{K}$

$= 2.2\ \mathrm{kJ} - 0.62\ \mathrm{kJ} = \underline{+1.6\ \mathrm{kJ}}$

25. $\Delta H = C_{p,\mathrm{m}}\Delta T$ [2.16]

$= 29.14\ \mathrm{J\ K^{-1}\ mol^{-1}} \times (37°\mathrm{C} - 15°\mathrm{C})$ [°C = K]

$= 29.14\ \mathrm{J\ K^{-1}\ mol^{-1}} \times 22\ \mathrm{K} = \underline{641\ \mathrm{J\ mol^{-1}}}$

$\Delta U = \Delta H - \Delta(pV) = \Delta H - \Delta(nRT) = \Delta H - R\,\Delta T$

$\Delta U = 641\ \mathrm{J\ mol^{-1}} - (8.31\ \mathrm{J\ K^{-1}\ mol^{-1}} \times 22\ \mathrm{K}) = \underline{458\ \mathrm{J\ mol^{-1}}}$

Thermochemistry

3

Assume all gases are perfect unless stated otherwise. All thermochemical data are for 298.15 K.

1. $\Delta_{fus}H^{\ominus} = 2.60\ kJ\ mol^{-1}$ [*Handbook of Chemistry and Physics*]

$$n = \frac{224 \times 10^3\ g}{22.99\ g\ mol^{-1}} = 9.74 \times 10^3\ mol$$

$$q = n\Delta_{fus}H^{\ominus} = 9.74 \times 10^3\ mol \times 2.60\ kJ\ mol^{-1}$$
$$= \underline{2.53 \times 10^4\ kJ}$$

3. The heat supplied to the sample is

$$q = I\mathcal{V}t \quad [2.4]$$
$$= 0.812\ A \times 11.5\ V \times 303\ s = 2.83 \times 10^3 J$$

$$q = \Delta H = n\Delta_{vap}H^{\ominus} \text{ (pressure is constant)}$$

$$n = \frac{427\ g}{60.04\ g\ mol^{-1}} = 0.0711\ mol$$

$$\Delta_{vap}H^{\ominus} = \frac{q}{n} = \frac{2.83 \times 10^3\ J}{0.0711\ mol} = 3.98 \times 10^4\ J\ mol^{-1} = \underline{39.8\ kJ\ mol^{-1}}$$

5. $n = \frac{100\ g\ ice}{18.0\ g\ mol} = 5.55\ mol$

The heat needed to melt 100 g of ice is

$$q_1 = n \times \Delta_{fus}H^{\ominus}$$
$$= 5.55\ mol \times 6.01\ kJ\ mol^{-1} = 33.4\ kJ$$

The heat needed to raise the temperature of the water from 0°C to 100°C is

$$q_2 = 100\ g \times 4.18\ J\ K^{-1}\ g^{-1} \times 100\ K = 4.18 \times 10^4\ J$$
$$= 41.8\ kJ$$

The heat needed to vaporize the water is

$q_3 = 5.55 \text{ mol} \times 40.7 \text{ kJ mol}^{-1} = 226 \text{ kJ}$

The total heat is

$q = q_1 + q_2 + q_3$

$\quad = 33.4 \text{ kJ} + 41.8 \text{ kJ} + 226 \text{ kJ} = 301 \text{ kJ}$

The graph of temperature against time is sketched in the figure below. Note that the length of the liquid + gas two phase line is longer than the solid + liquid line in proportion to their $\Delta_{trs}H$ values.

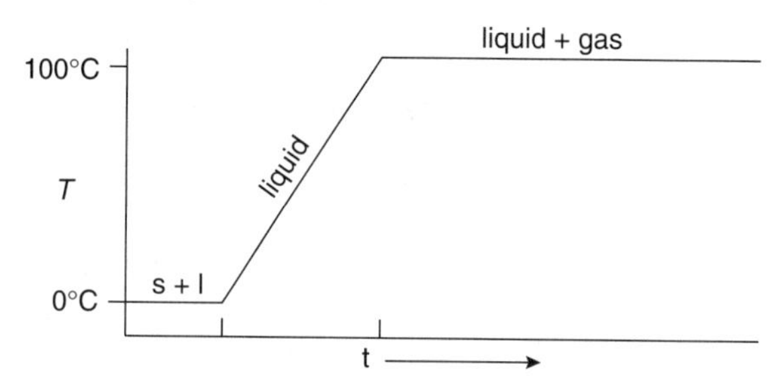

7. The mean bond enthalpy is the average of the three enthalpy changes given.

$$\Delta H_B \text{ (N – H)} = \left(\frac{460 + 390 + 314}{3}\right) \times \text{kJ mol}^{-1}$$

$$= \underline{388 \text{ kJ mol}^{-1}}$$

The bond dissociation energies and enthalpies refer to gas phase processes, and as a result of the dissociation the number of moles of gas phase particles increases. Therefore, because

$\Delta H = \Delta U + \Delta \nu_g RT$

ΔU_B (N – H) is expected to be <u>smaller</u> than ΔH_B (N – H)

9. (a) 1.00 mol N_2 consumed

$\Delta H^{\ominus} = 1.00 \text{ mol} \times (-92.22 \text{ kJ mol}^{-1}) = \underline{-92.22 \text{ kJ}}$

(b) 1.00 mol NH_3 formed

$$\Delta H^{\ominus} = \frac{1 \text{ mol N}_2}{2 \text{ mol NH}_3} \times 1 \text{ mol NH}_3 \times (-92.22 \text{ kJ mol}^{-1}) = \underline{-46.11 \text{ kJ}}$$

11. $C_6H_5C_2H_5(l) + \frac{21}{2} O_2(g) \rightarrow 8\ CO_2(g) + 5\ H_2O(l)$

$$\Delta_c H^{\ominus} = 8\ \Delta_f H^{\ominus} [CO_2(g)] + 5\ \Delta_f H^{\ominus} [H_2O(l)] - \Delta_f H^{\ominus} [C_6H_5C_2H_5(l)]$$

$$= 8 \times (-393.51\ \text{kJ mol}^{-1}) + 5 \times (-285.83\ \text{kJ mol}^{-1}) - (-12.5\ \text{kJ mol}^{-1})$$

$$= \underline{-4564.7\ \text{kJ mol}^{-1}}$$

13. $3\ C(s) + 3\ H_2(g) + O_2(g) \rightarrow CH_3COOCH_3(l) \qquad \Delta_f H^{\ominus} = -442\ \text{kJ mol}^{-1}$

$\Delta_f U^{\ominus} = \Delta_f H^{\ominus} - \Delta n_g RT,\ \Delta \nu_g = -4\ \text{mol}$

$\Delta_f U^{\ominus} = -442\ \text{kJ mol}^{-1} - (-4 \times 8.3145\ \text{J K}^{-1}\ \text{mol}^{-1} \times 298.15\ \text{K} \times 10^{-3}\ \text{kJ J}^{-1})$

$$= \underline{-432\ \text{kJ mol}^{-1}}$$

15. $q = n\Delta_c H^{\ominus}$ with $\Delta_c H^{\ominus} = -5157\ \text{kJ mol}^{-1}$ [Appendix 1]

Calculating q,

$$|q| = \frac{320 \times 10^{-3}\ \text{g}}{128.18\ \text{g mol}^{-1}} \times 5157\ \text{J mol}^{-1} = 12.87\ \text{kJ}$$

$$C = \frac{q}{\Delta T} = \frac{12.87\ \text{kJ}}{3.05\ \text{K}} = \underline{4.22\ \text{kJ K}^{-1}}$$

When phenol is used, $\Delta_c H^{\ominus} = -3054\ \text{kJ mol}^{-1}$ [Appendix 1]

$$|q| = \frac{100 \times 10^{-3}\text{g}}{94.12\ \text{g mol}^{-1}} \times 3054\ \text{J mol}^{-1} = 3.245\ \text{kJ}$$

$$\Delta T = \frac{q}{C} = \frac{3.245\ \text{kJ}}{4.22\ \text{kJ K}^{-1}} = \underline{0.769\ \text{K}}$$

17. (a) On the assumption that the calorimeter is a constant volume bomb calorimeter such as the one described in Figure 2.14, the heat released directly equals the internal energy of combustion, $\Delta_c U^{\ominus}$. Therefore, $\Delta_c U^{\ominus} = \underline{-1333\ \text{kJ mol}^{-1}}$.

(b) $HOOCCH{=}CHCOOH(s) + 3\ O_2(g) \rightarrow 4\ CO_2(g) + 2\ H_2O(l)$

$\Delta \nu_g = +1$

$\Delta_c H^{\ominus} = \Delta_c U^{\ominus} + \Delta \nu_g RT$

$$= -1333\ \text{kJ mol}^{-1} + (1\ \text{mol} \times 2.5\ \text{kJ mol}^{-1}) = \underline{-1331\ \text{kJ mol}^{-1}}$$

(c) $\Delta_c H^\ominus = 4\Delta_f H^\ominus [CO_2(g)] + 2\,\Delta_f H^\ominus [H_2O(l)] - \Delta_f H^\ominus$ (fumanic acid)

$= -1331 \text{ kJ mol}^{-1}$

$\Delta_f H^\ominus$ (fumanic acid) $= 4 \times (-393.51 \text{ kJ mol}^{-1}) + 2 \times (-285.83 \text{ kJ mol}^{-1}) + 1331 \text{ kJ mol}^{-1}$

$= \underline{-815 \text{ kJ mol}^{-1}}$

19. Because for $NH_3SO_2(s) \rightarrow NH_3(g) + SO_2(g)$ $\quad \Delta H^\ominus = +40 \text{ kJ mol}^{-1}$

for $NH_3(g) + SO_2(g) \rightarrow NH_3SO_2(s)$ $\quad \Delta H^\ominus = -40 \text{ kJ mol}^{-1}$

For the latter reaction

$\Delta_r H^\ominus = \Delta_f H^\ominus (NH_3SO_2) - \Delta_f H^\ominus (NH_3) - \Delta_f H^\ominus (SO_2) = -40 \text{ kJ mol}^{-1}$

Therefore, after solving for $\Delta_f H^\ominus (NH_3SO_2)$

$\Delta_f H^\ominus (NH_3SO_2, s) = \Delta_f H^\ominus (NH_3, g) + \Delta_f H^\ominus (SO_2, g) - 40 \text{ kJ mol}^{-1}$

$= (-46.11 - 296.83 - 40) \text{ kJ mol}^{-1} = \underline{-383 \text{ kJ mol}^{-1}}$

21. $\Delta_{trs} U^\ominus = w + q = -p_{ex}\,\Delta V + q$

$q = \Delta_{trs} H = +1.9 \text{ kJ mol}^{-1}$

$p_{ex} = 150 \text{ kbar} = 1.50 \times 10^5 \text{ bar} = 1.50 \times 10^{10} \text{ Pa}$

For 1 mol graphite

$$V_{gr} = \frac{12.01 \text{ g/mol}}{2.250 \text{ g/cm}^3} \times \frac{1 \text{ m}^3}{10^6 \text{ cm}^3} = 5.338 \times 10^{-6} \text{ m}^3 \text{ mol}^{-1}$$

For 1 mol diamond

$$V_{diam} = \frac{12.01 \text{ g/mol}}{3.510 \text{ g/cm}^3} \times \frac{1 \text{ m}^3}{10^6 \text{ cm}^3} = 3.422 \times 10^{-6} \text{ m}^3 \text{ mol}^{-1}$$

$\Delta V = V_{diam} - V_{gr} = 3.422 \times 10^{-6} \text{ m}^3 - 5.338 \times 10^{-6} \text{ m}^3$

$= -1.916 \times 10^{-6} \text{ m}^3 \text{ mol}^{-1}$

$-p_{ex}\,\Delta V = -1.50 \times 10^{10} \text{ Pa} \times (-1.916 \times 10^{-6} \text{ m}^3 \text{ mol}^{-1})$

$= 2.874 \times 10^4 \text{ J mol}^{-1} = 28.74 \text{ kJ mol}^{-1}$

$\Delta_{trs} U = 28.74 \text{ kJ mol}^{-1} + 1.9 \text{ kJ mol}^{-1} = \underline{+30.6 \text{ kJ mol}^{-1}}$

23. $C_3H_8(l) \rightarrow C_3H_8(g)$ $\quad \Delta_{vap} H^\ominus$

$C_3H_8(g) + 5\,O_2(g) \rightarrow 3\,CO_2(g) + 4\,H_2O(l)$ $\quad \Delta_c H^\ominus(g)$

(a) $\Delta_c H^\ominus(l) = \Delta_{vap} H^\ominus + \Delta_c H^\ominus(g)$

$= 15 \text{ kJ mol} - 2220 \text{ kJ mol}^{-1} = \underline{-2205 \text{ kJ mol}^{-1}}$

(b) $\Delta\nu_g = -2$ [5 O_2(g) replaced with 3 CO_2(g)]

$\Delta_c U^{\ominus}(\mathrm{l}) = \Delta_c H^{\ominus}(\mathrm{l}) - (-2)RT$

$= -2205\ \mathrm{kJ\ mol^{-1}} + (2 \times 2.5\ \mathrm{kJ\ mol^{-1}}) = \underline{-2200\ \mathrm{kJ\ mol^{-1}}}$

25. (a) $\Delta_r H^{\ominus} = \Delta_f H^{\ominus}(\mathrm{N_2O_4, g}) - 2\,\Delta_f H^{\ominus}(\mathrm{NO_2, g})$

$= [9.16 - 2 \times 33.18]\ \mathrm{kJ\ mol^{-1}} = \underline{-57.20\ \mathrm{kJ\ mol^{-1}}}$

(b) $\Delta_r H^{\ominus} = \frac{1}{2}\,\Delta_f H^{\ominus}(\mathrm{N_2O_4, g}) - \Delta_f H^{\ominus}(\mathrm{NO_2, g})$

$= \frac{1}{2}(9.16) - 33.18\ \mathrm{kJ\ mol^{-1}} = \underline{-28.6\ \mathrm{kJ\ mol^{-1}}}$

(c) $\Delta_r H^{\ominus} = 2 \times \Delta_f H^{\ominus}(\mathrm{HNO_3, aq}) + \Delta_f H^{\ominus}(\mathrm{NO, g}) - 3 \times \Delta_f H^{\ominus}(\mathrm{NO_2, g}) - \Delta_f H^{\ominus}(\mathrm{H_2O, l})$

$= [2 \times (-207.36) + 90.25 - 3 \times (33.18) - (-285.83)]\ \mathrm{kJ\ mol^{-1}}$

$= \underline{-138.2\ \mathrm{kJ\ mol^{-1}}}$

(d) $\Delta_r H^{\ominus} = \Delta_f H^{\ominus}(\text{propene, g}) - \Delta_f H^{\ominus}(\text{cyclopropane, g})$

$= [20.42 - 53.30]\ \mathrm{kJ\ mol^{-1}} = \underline{-32.88\ \mathrm{kJ\ mol^{-1}}}$

(e) In order to calculate $\Delta_r H^{\ominus}$ first write the net ionic equation:

$\mathrm{H^+(aq) + Cl^-(aq) + Na^+(aq) + OH^-(aq) \rightarrow Na^+(aq) + Cl^-(aq) + H_2O(l)}$

Simplifying we obtain

$\mathrm{H^+(aq) + OH^-(aq) \rightarrow H_2O(l)}$

$\Delta_r H^{\ominus} = \Delta_f H^{\ominus}(\mathrm{H_2O, l}) - \Delta_f H^{\ominus}(\mathrm{H^+, aq}) - \Delta_f H^{\ominus}(\mathrm{OH^-, aq})$

$= [-285.83 - 0 - (-229.99)]\ \mathrm{kJ\ mol^{-1}} = \underline{-55.84\ \mathrm{kJ\ mol^{-1}}}$

27. We use equations 3.5 and 3.6.

$\Delta_r H^{\ominus}(T_2) = \Delta_r H^{\ominus}[T_1] + \Delta_r C_p \Delta T$ [3.5]

$\Delta_r C_p = \sum \nu C_{p,m}(\text{prods}) - \sum \nu C_{p,m}(\text{reacs})$ [3.6]

$= (77.28 - 2 \times 37.20)\ \mathrm{J\ K^{-1}\ mol^{-1}} = +2.88\ \mathrm{J\ K^{-1}\ mol^{-1}}$

$\Delta_r H^{\ominus}(373\ \mathrm{K}) = \Delta_r H^{\ominus}(298\ \mathrm{K}) + \Delta_r C_p \Delta T$

$= -57.20\ \mathrm{kJ\ mol^{-1}} + (2.88\ \mathrm{J\ K^{-1}} \times 75\ \mathrm{K})$

$= (-57.20 + 0.22)\ \mathrm{kJ\ mol^{-1}} = \underline{-56.98\ \mathrm{kJ\ mol^{-1}}}$

29. (a) $\Delta_r C_p = 2 \times 9R - 3 \times \frac{7}{2}R = +\frac{15}{2}R$, <u>increase</u>

(c) $\Delta_r C_p = \frac{7}{2}R + (2 \times 9R) - 4R - (2 \times \frac{7}{2}R) = +\frac{21}{2}R$, <u>increase</u>

Thermodynamics: The Second Law

4

1. $\Delta S_{\text{sur}} = \dfrac{q_{\text{sur}}}{T}$ [4.7]

$$= \frac{120\ \text{J}}{293\ \text{K}} = \underline{0.410\ \text{J K}^{-1}}$$

3. $q = nC_{p,\text{m}}\Delta T$

$$q = \frac{1.25 \times 10^3\ \text{g}}{26.98\ \text{g mol}^{-1}} \times 24.35\ \text{J K}^{-1}\ \text{mol}^{-1} \times (-40\ \text{K})$$

$$= \underline{-45.1\ \text{kJ}}$$

$$\Delta S = C_p \ln \frac{T_2}{T_1} = nC_{p,\text{m}} \ln \frac{T_2}{T_1}$$

$$= \frac{1.25 \times 10^3\ \text{g}}{26.98\ \text{g mol}^{-1}} \times 24.35\ \text{J K}^{-1}\ \text{mol}^{-1} \times \ln \frac{260\ \text{K}}{300\ \text{K}} = \underline{-161\ \text{J K}^{-1}}$$

5. $\Delta S = nR \ln \dfrac{V_{\text{f}}}{V_{\text{i}}}$ [4.2] (Assume gas is perfect.)

$$= 8.31\ \text{J K}^{-1}\ \text{mol}^{-1} \times \ln\left(\frac{4.5\ \text{L}}{1.5\ \text{L}}\right) = +9.1\ \text{J K}^{-1}\ \text{mol}^{-1}$$

7. $\Delta S = nR \ln \dfrac{V_{\text{f}}}{V_{\text{i}}}$ [4.2]

Because entropy is a state function, it does not matter whether the change in state occurs reversibly or irreversibly. Therefore, for (a) and (b)

substitute $V = \dfrac{nRT}{p}$ into the expression for ΔS.

$$\Delta S = nR \ln \frac{p_{\text{i}}}{p_{\text{f}}}$$

$$= \frac{25\ \text{g}}{16.04\ \text{g mol}^{-1}} \times 8.315\ \text{J K}^{-1}\ \text{mol}^{-1} \times \ln \frac{185}{2.5} = \underline{56\ \text{J K}^{-1}}$$

9. Because entropy changes depend only on the initial and final states, it does not matter if the change is accomplished in one or more than one step. Therefore calculate the change in two steps.

ΔS for compression

$$\Delta S = nR\ln\frac{V_f}{V_i}\ [4.2]$$

$$= 1\ \text{mol} \times 8.31\ \text{J K}^{-1}\ \text{mol}^{-1} \times \ln\left(\frac{0.500\ \text{L}}{2.0\ \text{L}}\right) = -11.5\ \text{J K}^{-1}$$

ΔS for heating

$$\Delta S = nC_{V,m}\ln\frac{T_f}{T_i}\ [4.4],\ C_{V,m} = \frac{3}{2}R$$

$$= 1\ \text{mol} \times \frac{3}{2} \times 8.31\ \text{J K}^{-1}\ \text{mol}^{-1} \times \ln\left(\frac{400\ \text{K}}{300\ \text{K}}\right) = +3.59\ \text{J K}^{-1}$$

$$\Delta S_{total} = (-11.5 + 3.59)\ \text{J K}^{-1} = \underline{-7.9\ \text{J K}^{-1}}$$

11. Entropy changes occur in steps 1 and 3 and are the negatives of each other. Temperature changes in steps 2 and 4 are the negatives of each other.

Step 1: $\Delta S_1 = nR\ln\frac{V_2}{V_1} = +,\ \Delta T = 0$

Step 3: $\Delta S_3 = nR\ln\frac{V_1}{V_2} = -,\ \Delta T = 0$

Step 2: $\Delta S = 0,\ \Delta T_2 = -$

Step 4: $\Delta S = 0,\ \Delta T_4 = +$

$\Delta S_1 = -\Delta S_3,\ \Delta T_2 = -\Delta T_4$

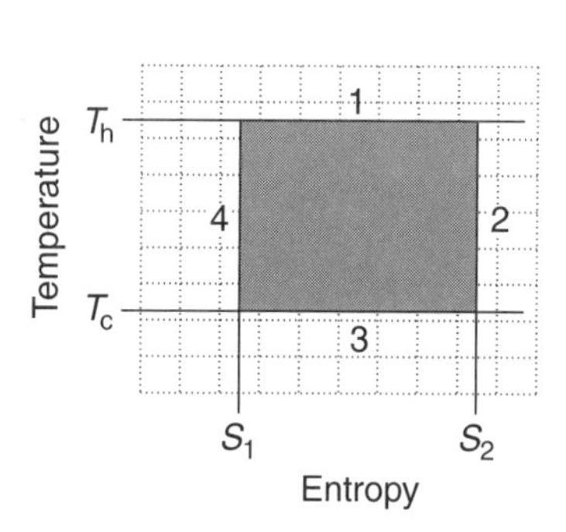

13. In a manner similar to equations 4.5 and 4.6, we may write in general,

$$\Delta_{trs}S = \frac{\Delta_{trs}H}{T_{trs}};\ \text{therefore}\ \Delta_{trs}S = \frac{+1.9\ \text{kJ mol}^{-1}}{2000\ \text{K}} = \underline{0.95\ \text{J K}^{-1}\ \text{mol}^{-1}}$$

15. (a) According to Trouton's rule, which applies well to hydrocarbons such as octane, $\Delta_{vap}S = \underline{+85\ \text{J K}^{-1}\ \text{mol}^{-1}}$

(b) $\Delta_{vap}S = \dfrac{\Delta_{vap}H}{T_b} = +85\ \text{J K}^{-1}\ \text{mol}^{-1}$

$$\Delta_{vap}H = \Delta_{vap}S \times T_b = +85\ \text{J K}^{-1}\ \text{mol}^{-1} \times 399\ \text{K} = \underline{+34\ \text{kJ K}^{-1}\ \text{mol}^{-1}}$$

17. (a) positive, due to greater disorder in the product, though the difference may not be large

    (b) negative, less disorder (smaller number of moles of gas) in the product

    (c) positive, two new substances are formed, resulting in greater disorder on the product side

19. $\Delta S = nC_{p,\mathrm{m}}\ln\dfrac{T_\mathrm{f}}{T_\mathrm{i}}$ for each substance in each reaction, therefore

    $\Delta_\mathrm{r}S = \Delta_\mathrm{r}C_p\ln\dfrac{T_\mathrm{f}}{T_\mathrm{i}}$ for each reaction

    (a) $\Delta_\mathrm{r}C_p = (2\text{ mol}\times 4R) - (3\text{ mol}\times\frac{7}{2}R) = -\frac{5}{2}R\text{ mol}$

    $$\Delta_\mathrm{r}S = -\frac{5}{2}R\text{ mol}\times\ln\left(\frac{283\text{ K}}{273\text{ K}}\right) = -0.75\text{ J K}^{-1}$$

    (b) $\Delta_\mathrm{r}C_p = (2\text{ mol}\times 4R) + (1\text{ mol}\times\frac{7}{2}R) - (1\text{ mol}\times 4R) - (2\text{ mol}\times\frac{7}{2}R)$

    $= +\frac{1}{2}R\text{ mol}$

    $$\Delta_\mathrm{r}S = +\frac{1}{2}R\text{ mol}\times\ln\left(\frac{283\text{ K}}{273\text{ K}}\right) = \underline{+0.15\text{ J K}^{-1}}$$

21. (a) $\Delta G = \Delta H - T\Delta S$ [4.14]

    $= -125\text{ kJ mol}^{-1} - 310\text{ K}\times(-126\text{ J K}^{-1}\text{ mol}^{-1}) = \underline{-86\text{ kJ mol}^{-1}}$

    (b) Yes, ΔG is negative.

    (c) $\Delta G = -T\Delta S_\mathrm{total}$ [4.15]

    $$\Delta S_\mathrm{total} = -\frac{\Delta G}{T} = -\left(\frac{-86\text{ kJ mol}^{-1}}{310\text{ K}}\right) = \underline{+0.28\text{ kJ K}^{-1}\text{ mol}^{-1}}$$

23. (a) Yes, coupling the two reactions can give a net ΔG that is negative, hence the overall process is spontaneous. For example, for one mole of glutamate and one mole of ATP, $\Delta G = (14.2 - 31)\text{ kJ mol}^{-1} = -17\text{ kJ mol}^{-1}$.

    (b) The minimum amount of ATP required is $\dfrac{1\text{ mol}\times(-14.2\text{ kJ mol}^{-1})}{-31\text{ kJ mol}^{-1}}$

    $= \underline{0.46\text{ mol ATP}}$

25. $n(\text{ATP}) = \dfrac{10^6}{6.02 \times 10^{23}\ \text{mol}^{-1}} = 1.7 \times 10^{-18}\ \text{mol}$

$\Delta G = 1.7 \times 10^{-18}\ \text{mol s}^{-1} \times (-31\ \text{kJ mol}^{-1}) = -5.3 \times 10^{-17}\ \text{kJ s}^{-1}$

$= -5.3 \times 10^{-14}\ \text{J s}^{-1}$

$$\text{Power density of cell} = \frac{\Delta G \text{ of cell per second}}{\text{volume of cell}}$$

$$V_{\text{cell}} = \frac{4}{3}\pi r^3 = \frac{4}{3}\pi(10 \times 10^{-6}\ \text{m})^3 = 4.2 \times 10^{-15}\ \text{m}^3$$

$$\text{Power density of cell} = \frac{5.3 \times 10^{-14}\ \text{J s}^{-1}}{4.2 \times 10^{-15}\ \text{m}^3} = \underline{13\ \text{W m}^{-3}}$$

$$\text{Power density of battery} = \frac{15\ \text{W}}{100\ \text{cm}^3 \times 10^{-6}\ \text{m}^3/\text{cm}^3} = \underline{150\ \text{kW m}^{-3}}$$

The battery has the greater power density.

Phase Equilibria: Pure Substances

5

1. The substance with the lower molar Gibbs energy is the more stable; therefore, <u>rhombic sulfur</u> is the more stable.

3. (a) We assume that the molar volume of water is approximately constant with respect to variation in pressure. Then

$$V_m = \frac{18.02 \text{ g mol}^{-1}}{1.03 \text{ g cm}^{-3}} = 17.5 \text{ cm}^3 \text{ mol}^{-1}$$

$$= 1.75 \times 10^{-5} \text{ m}^3 \text{ mol}^{-1}$$

$$\Delta p = g\rho h \text{ [0.2]} \quad [\Delta p = p_{\text{trench}} - p_{\text{surface}}]$$

$$= 9.81 \text{ m s}^{-2} \times 1.03 \text{ g cm}^{-3} \times \frac{1 \text{ kg}}{10^3 \text{ g}} \times \frac{10^6 \text{ cm}^3}{\text{m}^3} \times 11.5 \times 10^3 \text{ m}$$

$$= 1.16 \times 10^8 \text{ Pa} = 116 \text{ MPa}$$

$$\Delta G_m = V_m \Delta p = 1.75 \times 10^{-5} \text{ m}^3 \text{ mol}^{-1} \times 1.16 \times 10^8 \text{ Pa}$$

$$= 2.03 \times 10^3 \text{ J mol}^{-1} = \underline{+2.03 \text{ kJ mol}^{-1}}$$

(b) The pressure at the bottom of the mercury column is 1.000 atm by definition. 1.000 atm = 1.013×10^5 Pa.

$$\Delta p = 1.013 \times 10^5 \text{ Pa} - 0.160 \text{ Pa} \approx 1.013 \times 10^5 \text{ Pa}$$

$$V_m = \frac{200.6 \text{ g mol}^{-1}}{13.6 \text{ g cm}^3} = 14.8 \text{ cm}^3 \text{ mol} = 1.48 \times 10^{-5} \text{ m}^3 \text{ mol}^{-1}$$

$$\Delta G_m = V_m \Delta p = 1.48 \times 10^{-5} \text{ m}^3 \text{ mol}^{-1} \times 1.013 \times 10^5 \text{ Pa}$$

$$= \underline{+1.50 \text{ J mol}^{-1}}$$

5. $\Delta G_m = RT \ln \dfrac{p_f}{p_i}$ [5.4]

(a) $$\Delta G_m = 8.315 \text{ J K}^{-1} \text{ mol}^{-1} \times 293 \text{ K} \times \ln\left(\frac{2.0 \text{ bar}}{1.0 \text{ bar}}\right)$$

$$= 1.7 \times 10^3 \text{ J mol}^{-1} = \underline{+1.7 \text{ kJ mol}^{-1}}$$

(b) $$\Delta G_m = 8.315 \text{ J K}^{-1} \text{ mol}^{-1} \times 293 \text{ K} \times \ln\left(\frac{0.00027 \text{ atm}}{1.0 \text{ atm}}\right)$$

$$= -2.0 \times 10^4 \text{ J mol}^{-1} = \underline{-20 \text{ kJ mol}^{-1}}$$

7. (a) Attractive interactions tend to decrease the pressure of a gas relative to its perfect value for the same volume. We may qualitatively use equation 5.5 to decide that the molar Gibbs energy will be lowered relative to its "perfect" value.

   (b) Repulsive interactions have the opposite effect on the pressure of a gas, so we may qualitatively decide that they will raise the molar Gibbs energy relative to its "perfect" value.

9. For the transition S(rhombic) → S(monoclinic),

   $\Delta G_m = +0.33 \text{ kJ mol}^{-1}$

   $\Delta S_m = (32.6 - 31.8) \text{ J K}^{-1} \text{ mol}^{-1} = 0.8 \text{ J K}^{-1} \text{ mol}^{-1}$

   (b) $\Delta G_m = \Delta H_m - T\Delta S_m$

   We will assume that ΔH_m, and ΔS_m are roughly independent of temperature. We need $\Delta(\Delta G_m)$ to be $-0.33 \text{ kJ mol}^{-1}$ as a result of the change in temperature. That is

   $-(T_f - T_i)\Delta S_m = -0.33 \text{ kJ mol}^{-1}$

   Solve for T_f, $T_f = \dfrac{0.33 \times 10^3 \text{ J mol}^{-1}}{0.8 \text{ J K}^{-1} \text{ mol}^{-1}} + 298.15 \text{ K}$

   $T_f = \underline{7 \times 10^2 \text{ K}}$

11. The slope of a graph of G_m against T is $-S_m$, that is $\dfrac{dG_m}{dT} = -S_m$ [5.8b]

   The slopes in all phases are negative, because S_m is always positive, but

   $$\left|\frac{dG_m}{dT}(g)\right| > \left|\frac{dG_m}{dT}(l)\right| > \left|\frac{dG_m}{dT}(s)\right|$$

   because $S_m(g) > S_m(l) > S_m(s)$.

   Therefore, a graph of G_m against T appears as in the figure below. Absolute values of G_m are not known, but ΔG_m in each phase could be calculated as illustrated in Exercise 5.10.

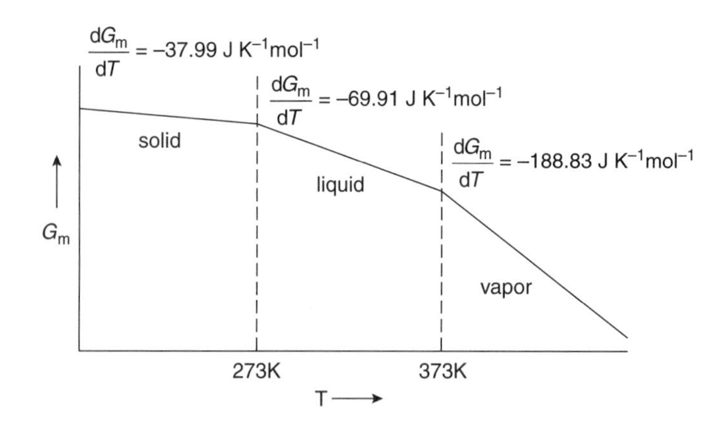

Note the discontinuous change in slopes at the transition temperatures.

13. (a) The Clapeyron equation for the solid–liquid phase boundary is

$$\frac{dp}{dT} = \frac{\Delta_{fus}S}{\Delta_{fus}V} \quad [5.7]$$

Using $\Delta_{fus}S = \dfrac{\Delta_{fus}H}{T_{trs}}$

The equation becomes

$$\frac{dp}{dT} = \frac{\Delta_{fus}H}{T_{fus}\Delta_{fus}V}$$

$$\Delta_{fus}V = V_m(l) - V_m(s) = M\left(\frac{1}{\rho_l} - \frac{1}{\rho_s}\right)$$

$$= 18.02\ \text{g mol}^{-1}\left(\frac{1}{0.99984\ \text{g cm}^{-3}} - \frac{1}{0.91671\ \text{g cm}^{-3}}\right)$$

$$= -1.634\ \text{cm}^3\ \text{mol}^{-1} = -1.634 \times 10^{-6}\ \text{m}^3\ \text{mol}^{-1}$$

$$\frac{dp}{dT} = \frac{6.008 \times 10^3\ \text{J mol}^{-1}}{273.15\ \text{K} \times (-1.634 \times 10^{-6}\ \text{m}^3\ \text{mol}^{-1})}$$

$$= -1.346 \times 10^7\ \text{Pa K}^{-1} = \underline{-134.6\ \text{bar K}^{-1}}$$

The slope is very steep!

(b) $\dfrac{\Delta p}{\Delta T} = -134.6\ \text{bar K}^{-1}$

For $\Delta T = -1$ K, $\Delta p = \underline{134.6\ \text{bar}}$

15. We use

$$\ln\frac{p'}{p} = \frac{\Delta_{vap}H}{R}\left(\frac{1}{T} - \frac{1}{T'}\right) \quad [5.9]$$

with $T = 293$ K, $p = 160$ m Pa, and $T' = 323$ K; then solve for p'.

$$\ln\frac{p'}{p} = \frac{59.30 \times 10^3\ \text{J mol}^{-1}}{8.315\ \text{J K}^{-1}\ \text{mol}^{-1}}\left(\frac{1}{293\ \text{K}} - \frac{1}{323\ \text{K}}\right) = 2.261$$

$$\frac{p'}{p} = 9.59$$

$p' = 9.59 \times 160$ m Pa $= 1.53 \times 10^3$ mPa $= \underline{1.53\ \text{Pa}}$

17. We use

$$\ln \frac{p'}{p} = \frac{\Delta_{vap}H}{R}\left(\frac{1}{T} - \frac{1}{T'}\right) \quad [5.9]$$

with $p' = 1$ atm $= 1.013 \times 10^5$ Pa, $p = 2.0 \times 10^4$ Pa, $T = 308$ K, $\Delta_{vap}H = 30.8$ kJ mol^{-1} [Table 3.1], and then solve for $T_b = T'$.

$$\ln\left(\frac{1.013 \times 10^5 \text{ Pa}}{2.0 \times 10^4 \text{ Pa}}\right) = \frac{30.8 \times 10^3 \text{ J mol}^{-1}}{8.315 \text{ J K}^{-1} \text{ mol}^{-1}}\left(\frac{1}{308} - \frac{1}{T_b}\right)$$

$$1.622 = 3704 \text{ K}\left(\frac{1}{308 \text{ K}} - \frac{1}{T_b}\right)$$

$T_b = \underline{356 \text{ K}}$

19. (a) The volume decreases as the vapor is cooled from 400 K, at constant pressure, in a manner described by the perfect gas equation

$$V = \frac{nRT}{p},$$

that is V is a linear function of T. This continues until 373 K is reached where the vapor condenses to a liquid and there is a large decrease in volume. As the temperature is lowered further to 273 K, liquid water freezes to ice. Only a small decrease in volume occurs in the liquid as temperature is decreased, and a small (~9%) increase in volume occurs when the liquid freezes. Water remains as a solid at 260 K.

(b) The cooling curve appears roughly as sketched in the figure below. The vapor and solid phases show a steeper rate of decline than for the liquid phase due to their smaller heat capacities. The temperature halt in the liquid plus vapor region is longer than for the liquid plus solid region due to its larger heat of transition.

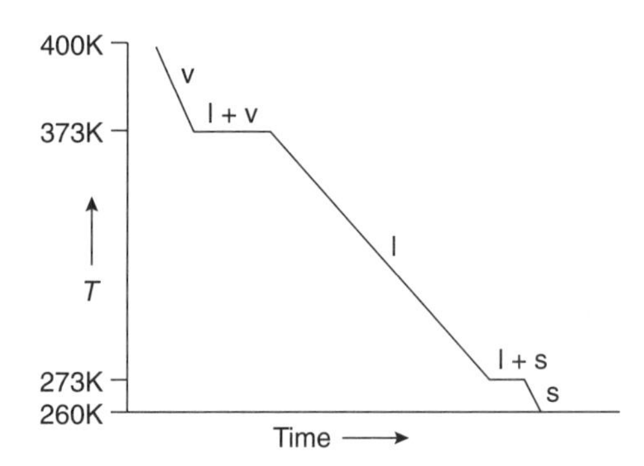

21. (a) The gaseous sample expands. (b) The sample contracts but remains gaseous because 320 K is greater than the critical temperature. (c) The gas contracts and forms a liquid-like substance without the appearance of a discernible surface. As the temperature lowers further to the solid phase boundary line, solid carbon dioxide forms in equilibrium with the liquid. At 210 K the sample has become all solid. (d) The solid expands slightly as the pressure is reduced and sublimes when the pressure reaches about 5 atm. (e) The gas expands as it is heated at constant pressure.

The Properties of Mixtures

1. $\text{mass} = 250.0\text{ mL} \times \dfrac{1\text{L}}{10^3\text{ mL}} \times \dfrac{0.112\text{ mol}}{1\text{L}} \times \dfrac{180.16\text{ g}}{\text{mol}} = \underline{5.04\text{ g}}$

3. $\text{mass} = 25.00\text{ mL}\ \dfrac{1\text{L}}{10^3\text{ mL}} \times \dfrac{0.245\text{ mol}}{\text{L}} \times \dfrac{75.07\text{ g}}{\text{mol}} = \underline{0.460\text{ g}}$

5. $x_{\text{sucrose}} = 0.124 = \dfrac{n_{\text{sucrose}}}{n_{\text{sucrose}} + n_{H_2O}} = \dfrac{m/M_{\text{sucrose}}}{m/M_{\text{sucrose}} + \dfrac{100\text{ g}}{18.02\text{ g mol}^{-1}}}$

   Solve the above equation for mass (m).

   $0.124\left(\dfrac{m}{M_{\text{sucrose}}} + 5.55\text{ mol}\right) = \dfrac{m}{M_{\text{sucrose}}}$

   $0.124 \times 5.55\text{ mol} = \dfrac{m}{M_{\text{sucrose}}}(1 - 0.124)$

   $m = \dfrac{0.124 \times 5.55\text{ mol} \times M_{\text{sucrose}}}{0.876}\ \left[M_{\text{sucrose}} = \dfrac{342.30\text{ g}}{\text{mol}}\right]$

   mass = $\underline{269\text{ g sucrose}}$

7. Let p = 1-propanol and b = 1-butanol

   $x_p = \dfrac{n_p}{n_p + n_b} \quad x_b = \dfrac{n_b}{n_p + n_b}$

   $n_p = \dfrac{40.0\text{ g}}{60.10\text{ g mol}^{-1}} = 0.665\text{ mol}$

   $n_b = \dfrac{60.0\text{ g}}{74.13\text{ g mol}^{-1}} = 0.809\text{ mol}$

   $x_p = \dfrac{0.665\text{ mol}}{0.665\text{ mol} \times 0.809\text{ mol}} = \underline{0.451}$

   $x_p + x_b = 1$, therefore $x_b = 1 - 0.451 = \underline{0.549}$

9. Let E = ethanol, W = water

$$n_E = 50.0\ \text{cm}^3 \times 0.789\ \text{g cm}^{-3} \times \frac{1\ \text{mol}}{46.07\ \text{g}} = 0.856\ \text{mol}$$

$$n_W = 50.0\ \text{cm}^3 \times 1.000\ \text{g cm}^{-3} \times \frac{1\ \text{mol}}{18.02\ \text{g}} = 2.775\ \text{mol}$$

$$x_E = \frac{n_E}{n_E + n_W} = \frac{0.856}{0.856 + 2.775} = 0.236$$

$$x_W = 1 - 0.236 = 0.764$$

From Figure 6.1 we may roughly estimate the partial molar volumes as $V_E = 54.5\ \text{cm}^3/\text{mol}$, $V_W = 17.9\ \text{cm}^3/\text{mol}$. Then

$$\begin{aligned} V &= n_E V_E + n_W V_W \\ &= 0.856\ \text{mol} \times 54.5\ \text{cm}^3\ \text{mol}^{-1} + 2.775\ \text{mol} \times 17.9\ \text{cm}^3\ \text{mol}^{-1} \\ &= \underline{96\ \text{cm}^3} \end{aligned}$$

11. $V = n_E V_E + n_W V_W = bV_E + n_W V_W$

We will assume that the solution contains 1.000 kg of water and solve for V_W.

$$n_W = \frac{1000\ \text{g}}{18.02\ \text{g mol}^{-1}} = 55.51\ \text{mol}$$

$$V_W = \frac{V - bV_E}{n_W}$$

$$bV_E/\text{mL} = 54.6664\, b - 0.72788\, b^2 + 0.084768\, b^3$$

$$\begin{aligned} (V - bV_E)/\text{mL} &= 1002.93 - 0.36394\, b^2 + 0.72788\, b^2 + 0.028256\, b^3 - 0.084768\, b^3 \\ &= 1002.93 + 0.36394\, b^2 - 0.056512\, b^3 \end{aligned}$$

$$V_W/\text{mL mol}^{-1} = \underline{18.067 + 6.556 \times 10^{-3}\, b^2 - 1.018 \times 10^{-3}\, b^3}$$

From an examination of the figure below, we see that the <u>maximum occurs at $b \approx 4.3\ \text{mol kg}^{-1}$</u> in agreement with the minimum in V_{ethanol}.

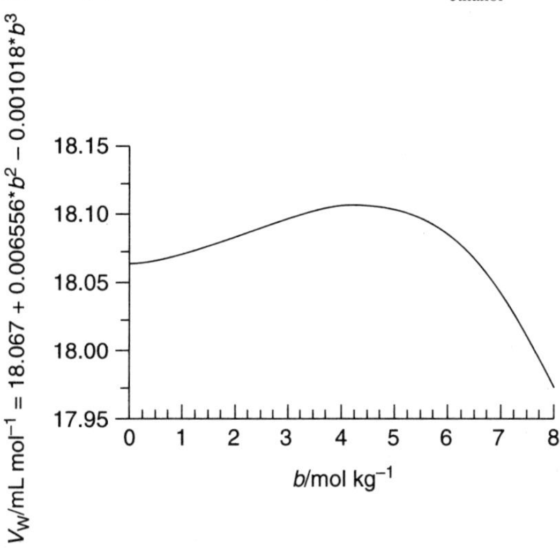

13. $\Delta G_m = RT(x_A \ln x_A + x_B \ln x_B + x_C \ln x_C)$

$A = N_2$, $B = O_2$, $C = Ar$

$\Delta G_m = 2.48\ \text{kJ mol}^{-1}\ (0.780 \ln 0.780 + 0.210 \ln 0.210 + 0.0096 \ln 0.0096)$

$= \underline{-1.40\ \text{kJ mol}^{-1}}$

Because the change in ΔG_m is negative upon the addition of argon, the mixing is spontaneous.

$\Delta S_m = -R(x_A \ln x_A + x_B \ln x_B + x_C \ln x_C)$

$= \underline{+4.71\ \text{J K}^{-1}\ \text{mol}^{-1}}$

15. Let us assume that a liter of seawater contains roughly 1000 g of water. Then

$$n_{\text{water}} = \frac{1000\ \text{g}}{18.02\ \text{g mol}^{-1}} = 55.5\ \text{mol}$$

$$n_{\text{solutes}} = 2 \times 0.50\ \text{mol} = 1.00\ \text{mol}$$

$$x_{\text{water}} = \frac{n_{\text{water}}}{n_{\text{water}} + n_{\text{solutes}}} = \frac{55.5}{56.5} = 0.982$$

$$p_{\text{water}} = x_{\text{water}}\, p^*_{\text{solutes}} = 0.982 \times 2.538\ \text{kPa} = \underline{2.49\ \text{kPa}}$$

17. $x_B = \dfrac{p_B}{K_B}$ [6.15]

Convert Torr to kPa. 760 Torr = 101.3 kPa

1 Torr = 0.1333 kPa

$$x_B = \frac{55\ \text{kPa}}{(8.6 \times 10^4\ \text{Torr}) \times 0.1333\ \text{kPa Torr}^{-1}} = \underline{4.8 \times 10^{-3}}$$

19. $K_{N_2} = 6.51 \times 10^7$ Torr and $K_{O_2} = 3.30 \times 10^7$ Torr. Therefore as in problem 4.18, the amount of dissolved gas in 1 kg of water is

$$n_{N_2} = \frac{10^3\ \text{g}}{18.02\ \text{g mol}^{-1}} \times \frac{p(N_2)}{6.51 \times 10^7\ \text{Torr}}$$

$$= 8.52 \times 10^{-7}\ \text{mol} \times (p/\text{Torr})$$

For $p_{N_2} = xp$ and $p = 760$ Torr

$n_{N_2} = (8.52 \times 10^{-7}\ \text{mol}) \times (x) \times (760) = 6.48x \times 10^{-4}\ \text{mol}$

and with $x = 0.78$,

$n_{N_2} = 0.78 \times 6.48 \times 10^{-4}\ \text{mol} = 5.1 \times 10^{-4}\ \text{mol} = 0.51\ \text{mmol}$

The molality of the solution is therefore approximately

$\underline{5.1 \times 10^{-4}\ \text{mol kg}^{-1}}$

For oxygen,

$$n_{O_2} = \frac{10^3\ \text{g}}{18.02\ \text{g mol}^{-1}} \times \frac{p_{O_2}}{3.30 \times 10^7\ \text{Torr}} = 1.68 \times 10^{-6}\ \text{mol} \times (p/\text{Torr})$$

For $p_{O_2} = xp$ and $p = 760$ Torr

$n_{O_2} = (1.68 \times 10^{-6}\ \text{mol}) \times (x) \times (760) = (1.28\ \text{mmol}) \times (x)$

when $x = 0.21$, $n_{O_2} = 0.27$ mmol. Hence the solution will be 0.27 mmol Kg^{-1} in O_2.

21. $p = p_A + p_B = xp_A^* + xp_B^* = xp_A^* + (1 - x_A)p_B^*$

Solving for x_A

$$x_A = \frac{p - p_B^*}{p_A^* - p_B^*}$$

For boiling under 0.50 atm (380 Torr) pressure, the combined vapor pressure must be 380 Torr, hence if A = toluene and B = *o*-xylene

$$x_A = \frac{380 - 150}{400 - 150} = 0.920,\ x_B = 0.080$$

The composition of the vapor is given by

$$y_A = \frac{p_A}{p} = \frac{x_A - p_A^*}{p_B^* + (p_A^* - p_B^*)x_A} = \frac{0.920 \times 400}{150 + [(400 - 150) \times 0.920]} = 0.968$$

and $y_B = 1 - 0.968 = 0.032$

23. Assume 150 cm^3 of water has a mass of 0.150 kg.

$$\Delta T = K_f b_B = 1.86\ \text{K kg mol}^{-1} \times \frac{7.5\ \text{g}}{342.3\ \text{g mol}^{-1} \times 0.150\ \text{kg}} = 0.27\ \text{K}$$

The freezing point will be approximately –0.27°C.

25. $K = \frac{[A_2]}{[A]^2}$, Initial [A] = n

$fn = [A_2]$ at equilibrium

$[A] = (1 - 2f)n$ and the total amount of solute is $(1 - f)n$. Therefore if the volume is V

$$K = \frac{fnV}{(1 - 2f)^2 n^2} = \frac{f}{(1 - 2f)^2 c} \quad \text{where } c = n/V$$

Vapor pressure, p is $p = x_{\text{solvent}}\, p^*$

$$p = x_s p^* = \frac{n_s p^*}{n_A + n_{A_2} + n_s} = \frac{n_s p^*}{(1-f)n + n_s}$$

$n_s = Vr$ with $r = \rho/M$, ρ = density of solvent

$p = \dfrac{rp^*}{(1-f)c + r}$ rearranging $f = 1 - \dfrac{r(p^* - p)}{cP}$ and, finally

$$K = \frac{1 - \dfrac{r(p^* - p)}{cp}}{c\left(1 - \dfrac{2r(p^* - p)}{cp}\right)^2}$$

27. $\pi V = n_B RT$ so that $\pi = \dfrac{mRT}{MV} = \dfrac{cRT}{M}$, $c = m/V$

$\pi = \rho g h$ [hydrostatic pressure] so

$$h = \left(\frac{RT}{\rho g M}\right) c$$

A plot of h versus c will have a slope of $\dfrac{RT}{\rho g M}$. The slope of this plot is 0.285

so $\dfrac{RT}{\rho g M} = \dfrac{0.285 \text{ cm}}{\text{gL}^{-1}} = 0.285 \times 10^{-2} \text{ m}^4 \text{ kg}^{-1}$

Therefore $M = \dfrac{RT}{\rho g \times 0.285 \times 10^{-2} \text{ m}^4 \text{ kg}^{-1}}$

$$= \frac{(8.315 \text{ JK}^{-1} \text{ mol}^{-1}) \times (298.15 \text{ K})}{(1.004 \times 10^3 \text{ kg m}^{-3}) \times (9.81 \text{ m s}^{-2}) \times (0.285 \times 10^{-2} \text{ m}^4 \text{ kg}^{-1})}$$

$= \underline{88.3 \text{ kg mol}^{-1}}$

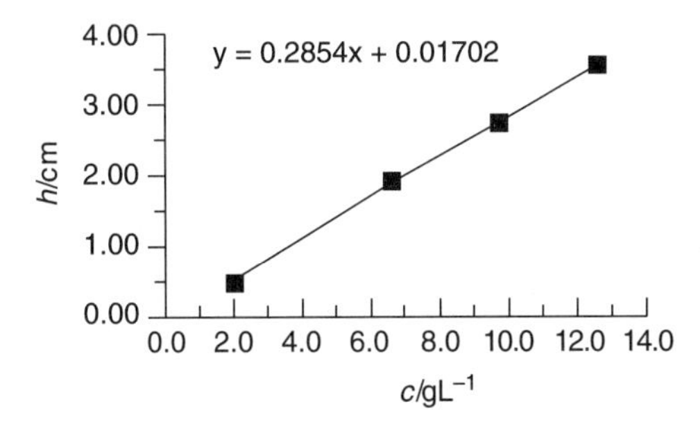

29. The data are plotted in the figure below. From tie line (a) on the graph, the vapor in equilibrium with liquid of composition $x_T = 0.250$ has $y_T = \underline{0.36}$. From tie line (b), for $x_O = 0.250$, $x_T = 0.750$, $y_T = \underline{0.81}$.

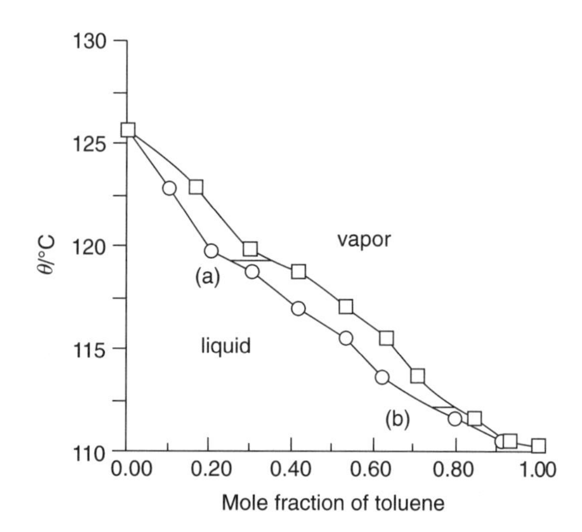

31. Refer to Figure 6.35 of the text. At b_3 there are two phases with compositions $x_A = 0.18$ and $x_A = 0.70$; their abundances are in the ratio 0.13 [lever rule]. Because C = 2 and P = 2 we have F = 2 (such as p and x). On heating, the phases merge, and the single-phase region is encountered. Then F = 3 (such as p, T, and x). The liquid comes into equilibrium with its vapor when the isopleth cuts the phase line. At this temperature, and for all points up to b_1, C = 2 and P = 2, implying that F = 2. The whole sample is a vapor above b_1.

33. The curves are shown in part (b) of the figure in the solution for 6.32. Note the eutectic halt for the isopleth b.

35. The phase diagram is sketched on page 35. (a) The mixture has a single liquid phase at all compositions. (b) When the composition reaches $x(C_6F_{10}) = 0.24$, the mixture separates into two liquid phases of composition $x = 0.24$ and 0.48. The relative amounts of the two phases change until the composition reaches x = 0.48. At all mole fractions greater than 0.48 in C_6F_{14} the mixture forms a single liquid phase.

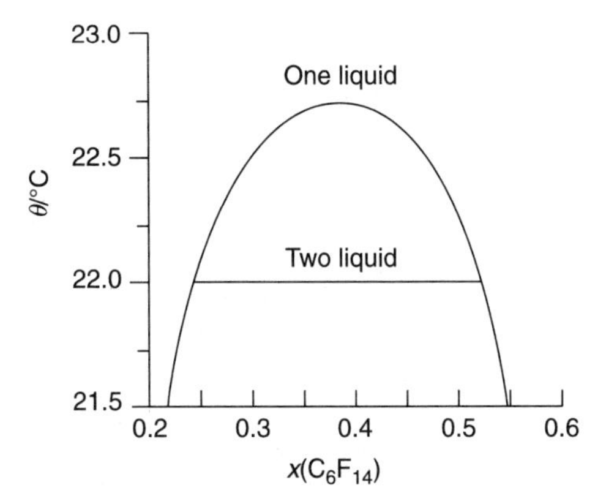

37. At roughly 34°C, solid begins to form as the state point enters the two-phase region. Within the two-phase region the proportion of liquid and solid can be determined by the lever rule. As the temperature is lowered through the two-phase region, the proportion of solid increases until, at roughly 20°C, the system becomes totally solid.

Principles of Chemical Equilibrium

7

1. (a) $Q = \dfrac{[G][Pi]}{[G6P]}$ H_2O is solvent

(b) $Q = \dfrac{[Gly-Ala]}{[Gly][Ala]}$ H_2O is solvent

(c) $Q = \dfrac{[MgATP^{2-}]}{[Mg^{2+}][ATP^{4-}]}$

(d) $Q = \dfrac{p^6_{CO_2}}{p^5_{O_2}[CH_3COCOOH]^2}$ H_2O is solvent

3. $\Delta_r G = \Delta_f G^{\ominus} + RT \ln Q$

for $\frac{1}{2}N_2(g) + \frac{3}{2}H_2(g) \rightarrow NH_3(g)$

$$Q = \frac{p_{NH_3}}{p^{1/2}_{N_2} p^{3/2}_{H_2}}$$

$$= \frac{4.0}{(3.0)^{1/2}(1.0)^{3/2}} = \frac{4.0}{\sqrt{3.0}}$$

Therefore

$$\Delta_r G = -16.45 \text{ kJ mol}^{-1} + RT \ln \frac{4.0}{\sqrt{3.0}}$$

$$= -16.45 \text{ kJ mol}^{-1} + 2.07 \text{ kJ mol}^{-1} = \underline{-14.38 \text{ kJ mol}^{-1}}$$

Because $\Delta_r G < 0$, the spontaneous direction of the reaction is toward the products.

5. $K_f = \dfrac{[C]}{[A][B]} = 0.224$

$$K_r = \frac{[A][B]}{[C]} = \frac{1}{K_f} = \frac{1}{0.224} = \underline{4.46}$$

7. $\Delta_r G^{\ominus} = -RT \ln K$ [7.8]

$$= -8.315 \text{ J K}^{-1} \text{ mol}^{-1} \times 400 \text{ K} \times \ln 2.07$$

$$= -2.42 \times 10^3 \text{ J mol}^{-1} = \underline{-2.42 \text{ kJ mol}^{-1}}$$

9. $\Delta_r G^\ominus = -RT \ln K$, therefore

$$\frac{\Delta_r G_1^\ominus}{\Delta_r G_2^\ominus} = \frac{-200 \text{ kJ mol}^{-1}}{-100 \text{ kJ mol}^{-1}} = \frac{-RT \ln K_1}{-RT \ln K_2} = 2$$

$2 \ln K_2 = \ln K_1$

$K_2^2 = K_1, K_2 = K_1^{1/2}$

$$\frac{K_1}{K_2} = \frac{K_1}{K_1^{1/2}} = \underline{K_1^{1/2}}$$

11. $\Delta_r G^\ominus = -RT \ln K = 0$

$\ln K = 0$

$K = \underline{1}$

13. $\text{ATP(aq)} + \text{H}_2\text{O(l)} \rightarrow \text{ADP(aq)} + \text{P}_\text{i}^-\text{(aq)} + \text{H}^+\text{(aq)}$

The standard value of the Gibbs energy quoted applies to the state where $a_{\text{H}^+} = 10^{-7}$ and all other activities are 1 (i.e., the biological standard state).

Hence $Q^\oplus = 1 \times 10^{-7}$

$\Delta_r G^\oplus = -30.5 \text{ kJ mol}^{-1} = \Delta_r G^\ominus + RT \ln Q^\oplus$

$$\Delta_r G = \Delta_r G^\ominus + \text{RT} \ln Q = \Delta_r G^\oplus + RT \ln\left(\frac{a_{\text{ADP}} a_{\text{Pi}^-}}{a_{\text{ATP}}}\right)$$

where we have used

$$Q = Q^\oplus \left(\frac{a_{\text{ADP}} a_{\text{Pi}^-}}{a_{\text{ATP}}}\right)$$

We assume that at these low concentrations we may replace activities with concentrations (all γ's = 1).

(a) $\Delta_r G = -30.5 \text{ kJ mol}^{-1} + 2.577 \text{ kJ mol}^{-1} \times \ln (1.0 \times 10^{-3})$

$= \underline{-48.3 \text{ kJ mol}^{-1}}$

(b) $\Delta_r G = -30.5 \text{ kJ mol}^{-1} + 2.577 \text{ kJ mol}^{-1} \times \ln (1.0 \times 10^{-6})$

$= \underline{-66.1 \text{ kJ mol}^{-1}}$

15. $\mu_\text{J} = \mu_\text{J}^\ominus + RT \ln a_\text{J}$ [7.3]

Assume $a_{\text{Na}^+} = [\text{Na}^+]$.

$\Delta G = \mu_{\text{Na}^+} \text{(outside)} - \mu_{\text{Na}^+} \text{(inside)}$

$$= RT \ln (140) - RT \ln (10) = RT \ln \left(\frac{140}{10}\right)$$

$= 8.315 \text{ J K}^{-1} \text{ mol}^{-1} \times 310 \text{ K} \times \ln 14$

$= 6.8 \times 10^3 \text{ J mol}^{-1} = \underline{6.8 \text{ kJ mol}^{-1}}$

17. We use the van't Hoff equation [7.14]

$$\ln K' - \ln K = \frac{\Delta_r H^{\ominus}}{R}\left(\frac{1}{T} - \frac{1}{T'}\right)$$

Substituting $\ln K = -\frac{\Delta_r G^{\ominus}}{RT}$ we obtain

$$\frac{\Delta_r G^{\ominus}}{T'} - \frac{\Delta_r G^{\ominus}}{T} = \Delta_r H^{\ominus}\left(\frac{1}{T'} - \frac{1}{T}\right)$$

$T = 1280$ K, T' = temperature at which $K' = 1$. When $K' = 1$, $\ln K' = 0$, and $\Delta_r G^{\ominus\prime} = 0$. This occurs when

$$-\frac{\Delta_r G^{\ominus}}{T} = \Delta_r H^{\ominus}\left(\frac{1}{T'} - \frac{1}{T}\right)$$

We solve for T′

$$\frac{1}{T'} = \frac{1}{T} - \frac{\Delta_r G^{\ominus}}{T\Delta_r H^{\ominus}} = \frac{1}{T}\left(1 - \frac{\Delta_r G^{\ominus}}{\Delta_r H^{\ominus}}\right)$$

$$= \frac{1}{1280\ \text{K}}\left(1 - \frac{33\ \text{kJ mol}^{-1}}{224\ \text{kJ mol}^{-1}}\right) = 6.66 \times 10^{-4}\ \text{K}^{-1}$$

$T = \underline{1.5 \times 10^3\ \text{K}}$

19. We look up $\Delta_f G^{\ominus}$ for each compound and note the sign.
(a) –, exergonic, (b) +, endergonic, (c) +, endergonic (d) –, exergonic

21. In each case, calculate $\Delta_r G^{\ominus}$ from the values of $\Delta_f G^{\ominus}$ found in Appendix 1. Then from $\Delta_r G^{\ominus} = -RT \ln K$ decide which have $K > 1$. If $\ln K > 0$, $K > 1$. $\ln K > 0$, if $\Delta_r G^{\ominus} < 0$. So if $\Delta_r G^{\ominus} < 0$, $K > 1$.

(a) $\Delta_r G^{\ominus} = 2\,\Delta_f G^{\ominus}(CH_3COOH, l) - 2\,\Delta_f G^{\ominus}(CH_3CHO, g)$

$= [2 \times (-389.9) - 2 \times (-128.86)]$ kJ mol^{-1}

$= \underline{-522.1\ \text{kJ mol}^{-1}}$, $\underline{K > 1}$

(b) $\Delta_r G^{\ominus} = 2\,\Delta_f G^{\ominus}(AgBr, s) - 2\,\Delta_f G^{\ominus}(AgCl, s)$

$= [2 \times (-96.90) - 2 \times (-109.79)]$ kJ mol^{-1}

$= \underline{+25.78\ \text{kJ mol}^{-1}}$, $\underline{K < 1}$

(c) $\Delta_r G^{\ominus} = \Delta_f G^{\ominus}(HgCl_2, s) = \underline{-178.6\ \text{kJ mol}^{-1}}$, $\underline{K > 1}$

(d) $\Delta_r G^{\ominus} = \Delta_f G^{\ominus}(Zn^{2+}, aq) - \Delta_f G^{\ominus}(Cu^{2+}, aq)$

$= [-147.06 - 65.49]$ kJ mol^{-1} $= \underline{-212.55\ \text{kJ mol}^{-1}}$, $\underline{K > 1}$

(e) $\Delta_r G^{\ominus} = 12\,\Delta_f G^{\ominus}(CO_2, g) + 11\,\Delta_f G^{\ominus}(H_2O, l) - \Delta_f G^{\ominus}(C_{12}H_{22}O_{11}, s)$

$= [12 \times (-394.36) + 11 \times (-237.13) - (-1543)]$ kJ mol^{-1}

$= \underline{-5798\ \text{kJ mol}^{-1}}$, $\underline{K > 1}$

23. We will see in Chapter 9 that $\Delta_r G^\ominus = \nu FE^\ominus$; therefore if $\Delta_r G^\ominus$ for a reaction is negative, it can in principle be used as the basis of a fuel cell, as a negative $\Delta_r G^\ominus$ results in a positive cell voltage, $E^\ominus$

From Appendix 1, $\Delta_f G^\ominus(\mathrm{NH_3, g}) = -16.5\ \mathrm{kJ\ mol^{-1}}$

$$\text{For 100 g, } n = 100\ \mathrm{g\ N_2} \times \frac{1\ \mathrm{mol}}{28.02\ \mathrm{g}} \times \frac{2\ \mathrm{mol\ NH_3}}{1\ \mathrm{mol\ N_2}} = 7.14\ \mathrm{mol\ NH_3}$$

$$\begin{aligned} w(\max) = \Delta G^\ominus &= n \times \Delta_r G^\ominus \\ &= 7.14\ \mathrm{mol} \times (-16.5\ \mathrm{kJ\ mol^{-1}}) \\ &= \underline{-118\ \mathrm{kJ}} \end{aligned}$$

25. In order to answer the question about energy effectiveness we need to compare $\Delta_r G^\ominus$ for the combustion of sucrose and glucose. The value for 1.0 kg of glucose was determined to be -1.57×10^4 kJ in Exercise 3.24. Here we calculate $\Delta_r G^\ominus$ for the combustion of sucrose.

(a) water vapor

$$\mathrm{C_{12}H_{22}O_{11}(s) + 12\ O_2(g) \rightarrow 12\ CO_2\ (g) + 11\ H_2O\ (g)}$$

$$\begin{aligned} \Delta_r G^\ominus &= 12 \times \Delta_f G^\ominus(\mathrm{CO_2, g}) + 11 \times \Delta_f G^\ominus(\mathrm{H_2O, g}) - \Delta_f G^\ominus(\text{sucrose}) \\ &= [12 \times (-394.36) + 11 \times (-228.57) - (-1543)]\ \mathrm{kJ\ mol^{-1}} \\ &= -5704\ \mathrm{kJ\ mol^{-1}} \end{aligned}$$

$$n\ (\text{sucrose}) = \frac{1000\ \mathrm{g}}{342.30\ \mathrm{g\ mol^{-1}}} = 2.92\ \mathrm{mol}$$

$$n \times \Delta_r G^\ominus = 2.92\ \mathrm{mol} \times (-5704\ \mathrm{kJ\ mol^{-1}}) = -1.67 \times 10^4\ \mathrm{kJ}$$

non-expansion work = $\underline{1.67 \times 10^4\ \mathrm{kJ}}$

expansion work, $w = -p_{cx}\Delta V \quad p = p_{ex} = 1.00$ atm

$$\Delta V = \frac{RT}{p}\Delta\nu_g \qquad \Delta\nu_g = 2.92\ \mathrm{mol} \times 11 = 31.9\ \mathrm{mol}$$

$$\begin{aligned} w = -RT\,\Delta\nu_g &= -2.478\ \mathrm{kJ\ mol^{-1}} \times 31.9\ \mathrm{mol} \\ &= -79.6\ \mathrm{kJ} \\ &= 79.6\ \mathrm{kJ\ expansion\ work\ done} \end{aligned}$$

$$\begin{aligned} \text{total work done} &= 1.67 \times 10^4\ \mathrm{kJ} + 79.0\ \mathrm{kJ} \\ &= \underline{1.68 \times 10^4\ \mathrm{kJ}} \end{aligned}$$

(b) liquid water.

From Exercise 3.21(e), $\Delta_r G^{\ominus} = -5798$ kJ mol^{-1}

non-expansion work = $\underline{1.69 \times 10^4 \text{ kJ}}$

expansion work = $\underline{0}$ ($\Delta\nu_g = 0$)

total work = $\underline{1.69 \times 10^4 \text{ kJ}}$

27. To calculate the standard free energy of formation, we need to calculate the entropy of formation and the enthalpy of formation.

$$6C(s) + 3H_2(g) + \frac{1}{2}O_2(g) \rightarrow C_6H_5OH(s)$$

$$\Delta_f G^{\ominus} = \Delta_f H^{\ominus} - T\Delta_f S^{\ominus}$$

$$\Delta_f S^{\ominus} = S_m^{\ominus}(C_6H_5OH, s) - 6\, S_m^{\ominus}(C, s) - 3\, S_m^{\ominus}(H_2, g) - \frac{1}{2}S_m^{\ominus}(O_2, g)$$

$$= (144.0 - 6 \times 5.740 - 3 \times 130.68 - \frac{1}{2} \times 205.14)\ \text{J K}^{-1}\ \text{mol}^{-1}$$

$$= -385.05\ \text{J K}^{-1}\ \text{mol}^{-1}$$

$$C_6H_5OH(s) + 7O_2(g) \rightarrow 6CO_2(g) + 3H_2O(l)$$

$$\Delta_r H^{\ominus} = 6\,\Delta_f H^{\ominus}(CO_2, g) + 3\,\Delta_f H^{\ominus}(H_2O, l) - \Delta_f H^{\ominus}(C_6H_5OH, s)$$

Rearranging

$$\Delta_f H^{\ominus}(C_6H_5OH, s) = 6\,\Delta_f H^{\ominus}(CO_2, g) + 3\,\Delta_f H^{\ominus}(H_2O, l) - \Delta_c H^{\ominus}$$

$$= [6 \times (-393.51) + 3(-285.83) - (-3054)]\ \text{kJ mol}^{-1}$$

$$= -164.55\ \text{kJ mol}^{-1}$$

Therefore

$$\Delta_f G^{\ominus} = -164.55\ \text{kJ mol}^{-1} - 298.15\ \text{K} \times (-0.38505\ \text{kJ K}^{-1}\ \text{mol}^{-1})$$

$$= \underline{-49.8\ \text{kJ mol}^{-1}}$$

29. $CH_3COCO_2^-(aq) \rightarrow CH_3CHO(g) + CO_2(g)$

$$\Delta_r G^{\ominus} = \Delta_f^{\ominus}(CH_3CHO, g) + \Delta_f G^{\ominus}(CO_2, g) - \Delta_f G^{\ominus}(CH_3COCO_2^-, g)$$

$$= (-133 - 394.36 + 474)\ \text{kJ mol}^{-1} = \underline{-53\ \text{kJ mol}^{-1}}$$

31. $\Delta_r G$ values do not depend on the presence or absence of a catalyst, so the answer is <u>the same</u> as in the solution to Exercise 7.29.

33. The reaction is

$AMP(aq) + H_2O(l) \rightarrow A(aq) + P_i^-(aq) + H^+(aq)$

Therefore, as in Example 7.3 in the text, the relationship between the standard state values of $\Delta_r G$ is

$$\Delta_r G^{\ominus} = \Delta_r G^{\oplus} - RT \ln (1 \times 10^{-7})$$

$$= -14 \text{ kJ mol}^{-1} - [8.31 \times 10^{-3} \text{ kJ K}^{-1} \text{ mol}^{-1} \times 298 \text{ K} \times (-16.1 \times 2)]$$

$$= \underline{+26 \text{ kJ mol}^{-1}}$$

35. Let B = borneol, I = isoborneol

For B $\rightarrow$ I

$$Q = \frac{p_I}{p_B}$$

$$p_B = x_B p$$

$$= \frac{0.15 \text{ mol}}{0.15 \text{ mol} + 0.30 \text{ mol}} \times 600 \text{ Torr} = 200 \text{ Torr}$$

$$p_I = p - p_B = 400 \text{ torr}$$

$$Q = \frac{p_I}{p_B} = 2.00$$

$$\Delta_r G = \Delta_r G^{\ominus} + RT \ln Q$$

$$= +9.4 \text{ kJ mol}^{-1} + (8.315 \text{ J K}^{-1} \text{ mol}^{-1} \times 503 \text{ K} \times \ln 2.00)$$

$$= \underline{+12.3 \text{ kJ mol}^{-1}}$$

37. $N_2(g) \quad + \quad 3H_2(g) \quad \rightleftharpoons \quad 2NH_3(g)$

$(1.00 - x) \qquad (4.00 - 3x) \qquad 2x \qquad$ partial pressure at equilibrium

$$K = \frac{p_{NH_3}^2}{p_{N_2} p_{H_2}^3} = \frac{(2x)^2}{(1.00 - x)(4.00 - 3x)^3} = 89.8$$

The expression above needs to be expanded and the resulting equation solved for x.

$$2424.6x^4 - 12123x^3 + 22625.6x^2 - 18678.4x + 5747.2 = 0$$

This equation is most readily solved with mathematical software on a computer. With the use of Math Cad 8, the root found is 0.968.

$$p_{NH_3} = 2 \times 0.968 = 1.936 \text{ bar}$$

$$p_{N_2} = 1.00 - 0.968 = 0.032 \text{ bar}$$

$p_{H_2} = 4.00 - (3 \times 0.968) = 1.096$ bar

Total $p = 3.064$ bar

$$x_{NH_3} = \frac{1.936}{3.064} = \underline{0.632}$$

$$x_{N_2} = \frac{0.032}{3.064} = \underline{0.010}$$

$$x_{H_2} = \frac{1.096}{3.064} = \underline{0.358}$$

39. $SbCl_5(g) \rightleftharpoons SbCl_3(g) + Cl_2(g)$

$$K = \frac{p_{SbCl_3} p_{Cl_2}}{p_{SbCl_5}} = 3.5 \times 10^{-4}$$

$$= \frac{(0.20) \cdot p_{Cl_2}}{(0.15)} = 3.5 \times 10^{-4}$$

$p_{Cl_2} = \underline{2.6 \times 10^{-4} \text{ bar}}$

41.

	$N_2(g)$	+	$3H_2(g)$	$\rightleftharpoons$	$2NH_3(g)$
Initial/atm	0.020		0.020		0
Change	$-x$		$-3x$		$+2x$
Equil.	$0.020 - x$		$0.020 - 3x$		$2x$

$$K = \frac{p^2_{NH_3}}{p_{N_2} p^2_{NH_3}} = 0.036$$

Because at this temperature $K << 1$, we may, at least initially make the approximation that x and $3x$ are small compared to 0.020. Then in a second approximation we may get a better value.

$$0.036 = \frac{4x^2}{(0.02)(0.02)^3}$$

Solving for x yields

$x = 3.8 \times 10^{-5}$

$$\text{Then } 0.036 = \frac{4x^2}{(0.019962) \times (0.019886)^3}$$

Solving again for x yields

$x = 3.8 \times 10^{-5}$

$2x = p_{NH_3} = \underline{7.6 \times 10^{-5}\text{ bar}}$

$p_{N_2} = 0.020\text{ bar} - 3.8 \times 10^{-5}\text{ bar} \cong \underline{0.020\text{ bar}}$

$p_{H_2} = 0.020\text{ bar} - 3 \times 3.8 \times 10^{-5}\text{ bar} \cong \underline{0.020\text{ bar}}$

43. $U(s) + \frac{3}{2} H_2(g) \rightleftharpoons UH_3(s)$

$K = \frac{1}{p_{H_2}^{3/2}}$ $\quad p_{H_2} = p$ Pressures here are unitless pressures relative to the standard pressure or 1 bar.

$\ln K = \ln p^{-3/2} = -\frac{3}{2} \ln p$

$\Delta_r G^{\ominus} = -RT \ln K = \frac{3}{2} \ln p$

$= \frac{3}{2} \times 8.315\text{ J K}^{-1}\text{ mol}^{-1} \times 500\text{ K} \times \ln \frac{1.04\text{ Torr}}{750\text{ Torr bar}^{-1}}$

$= \underline{-41.0\text{ kJ mol}^{-1}}$

45. $\ln \frac{K'}{K} = \frac{\Delta_r H^{\ominus}}{R}\left(\frac{1}{T} - \frac{1}{T'}\right)$

Solving for $\Delta_r H^{\ominus}$

$$\Delta_r H^{\ominus} = \frac{R\ln(K'/K)}{\left(\frac{1}{T} - \frac{1}{T'}\right)}$$

$T' = 308$ K, hence with $K'/K = k$

$$\Delta_r H^{\ominus} = \frac{8.315\text{ J K}^{-1}\text{ mol}^{-1} \times \ln k}{\left(\frac{1}{298\text{ K}} - \frac{1}{308\text{ K}}\right)} = 76\text{ kJ mol}^{-1} \times \ln k$$

Therefore

(a) $k = 2$, $\Delta_r H^{\ominus} = 76\text{ kJ mol}^{-1} \times \ln 2 = \underline{+53\text{ kJ mol}^{-1}}$

(b) $k = 1/2$, $\Delta_r H^{\ominus} = 76\text{ kJ mol}^{-1} \times \ln 1/2 = \underline{-53\text{ kJ mol}^{-1}}$

47. We first calculate K at 25°C.

$$\ln K = -\frac{\Delta_r G^{\ominus}}{RT} = \frac{+880 \text{ kcal mol}^{-1}}{1.987 \times 10^{-3} \text{ kcal K}^{-1} \text{ mol}^{-1} \times 298 \text{ K}}$$

$$= 1486$$

$K = e^{1486} = 10^{645}$ which is infinite for all practical purposes.

Then we use

$$\ln K' = \ln K + \frac{\Delta_r H^{\ominus}}{R}\left(\frac{1}{T} - \frac{1}{T'}\right)$$

$$= 1486 + \frac{3560 \text{ kcal mol}^{-1}}{1.987 \times 10^{-3} \text{ kcal K}^{-1} \text{ mol}^{-1}}\left(\frac{1}{298 \text{ K}} - \frac{1}{310 \text{ K}}\right)$$

$$= 1486 + 232 = 1718$$

Hence

$K' = e^{1718} = \underline{10^{746}}$ which is even more infinite.

Consequences of Equilibrium

1. (a)

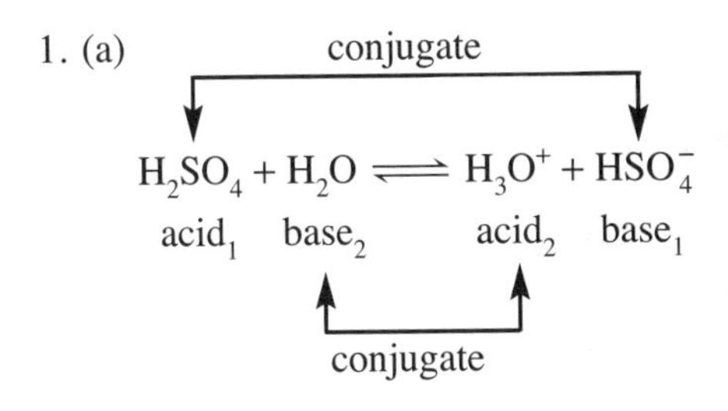

(b)

$$\underset{\text{acid}_1}{HF} + \underset{\text{base}_2}{H_2O} \rightleftharpoons \underset{\text{acid}_2}{H_3O^+} + \underset{\text{base}_1}{F^-}$$

conjugate: HF / F^- (acid$_1$ / base$_1$); conjugate: H_2O / H_3O^+ (base$_2$ / acid$_2$)

(c)

$$\underset{\text{acid}_1}{C_6H_5NH_3^+} + \underset{\text{base}_2}{H_2O} \rightleftharpoons \underset{\text{acid}_2}{H_3O^+} + \underset{\text{base}_1}{C_6H_5NH_2}$$

conjugate: $C_6H_5NH_3^+$ / $C_6H_5NH_2$ (acid$_1$ / base$_1$); conjugate: H_2O / H_3O^+ (base$_2$ / acid$_2$)

(d)

$$\underset{\text{acid}_1}{H_2PO_4^-} + \underset{\text{base}_2}{H_2O} \rightleftharpoons \underset{\text{acid}_2}{H_3O^+} + \underset{\text{base}_1}{HPO_4^{2-}}$$

conjugate: $H_2PO_4^-$ / HPO_4^{2-} (acid$_1$ / base$_1$); conjugate: H_2O / H_3O^+ (base$_2$ / acid$_2$)

(e) conjugate (HCOOH ↔ HCO_2^-)

$$\underset{\text{acid}_1}{HCOOH} + \underset{\text{base}_2}{H_2O} \rightleftharpoons \underset{\text{acid}_2}{H_3O^+} + \underset{\text{base}_1}{HCO_2^-}$$

conjugate (H_2O ↔ H_3O^+)

(f) conjugate ($NH_2NH_3^+$ ↔ NH_2NH_2)

$$\underset{\text{acid}_1}{NH_2NH_3^+} + \underset{\text{base}_2}{H_2O} \rightleftharpoons \underset{\text{acid}_2}{H_3O^+} + \underset{\text{base}_1}{NH_2NH_2}$$

conjugate (H_2O ↔ H_3O^+)

3. (a) $K_w = 2.5 \times 10^{-14} = [H_3O^+][OH^-] = x^2$

$[H_3O^+] = \sqrt{2.5 \times 10^{-14}} = 1.6 \times 10^{-7}$ mol L^{-1}

pH = $-\log[H_3O^+]$ = $\underline{6.80}$

(b) $[OH^-] = [H_3O^+] = 1.6 \times 10^{-7}$ mol L^{-1}

pOH = $-\log[OH^-]$ = $\underline{6.80}$

5. (a) pH = $-\log(1.5 \times 10^{-5})$ = $\underline{4.82}$ pOH = 14.00 – pH = 14.00 – 4.82 = $\underline{9.18}$

(b) pH = $-\log(1.5 \times 10^{-3})$ = $\underline{2.82}$ pOH = 14.00 – 2.28 = $\underline{11.18}$

(c) pH = $-\log(5.1 \times 10^{-14})$ = $\underline{13.29}$ pOH = 14.00 – 13.29 = $\underline{0.71}$

(d) pH = $-\log(5.01 \times 10^{-5})$ = $\underline{4.30}$ pOH = 14.00 – 4.30 = $\underline{9.70}$

7. The general rule is:

(1) salt of strong acid and strong base is neutral.

(2) salt of strong acid and weak base is acidic.

(3) salt of weak acid and strong base is basic.

(4) salt of weak acid and weak base is often close to neutral.

(a) acidic; $NH_4^+(aq) + H_2O(l) \rightleftharpoons H_3O^+(aq) + NH_3(aq)$

(b) basic; $H_2O(l) + CO_3^{2-}(aq) \rightleftharpoons HCO_3^-(aq) + OH^-(aq)$

(c) basic; $H_2O(l) + F^-(aq) \rightleftharpoons HF(aq) + OH^-(aq)$

(d) neutral

(e) acidic; $[Al(H_2O)_6]^{3+}(aq) + H_2O(l) \rightleftharpoons [Al(H_2O)_5OH]^{2+}(aq) + H_3O^+(aq)$

(f) acidic; $[Co(H_2O)_6]^{2+}(aq) + H_2O(l) \rightleftharpoons [Co(H_2O)_5OH]^+(aq) + H_3O^+(aq)$

9. (a) $K_a = \dfrac{[H_3O^+][L^-]}{[HL]}$ HL = lactic acid

if $[L^-] = [HL]$ then $K_a = [H_3O^+]$

$-\log K_a = -\log [H_3O^+]$

$pK_a = pH = 3.08$

$K_a = \underline{8.3 \times 10^{-4}}$

(b)

	$HL(aq)$	+ $H_2O(l)$	$\rightleftharpoons$ $H_3O^+(aq)$	+ $L^-(aq)$
Initial	x	—	—	$2x$
Change	$-y$	—	$+y$	$+y$

$$K_a = \frac{[H_3O^+][L^-]}{[HL]} = \frac{[y][y+x]}{[2x-y]} \approx \frac{[y][x]}{[2x]} = 8.3 \times 10^{-4}$$

$y = 2(8.3 \times 10^{-4}) = 1.66 \times 10^{-3}\ \text{mol L}^{-1}$
$= [H_3O^+]$

$pH = \underline{2.78}$

11. (a)

	C_6H_5COOH	+ H_2O	$\rightleftharpoons$ H_3O^+	+ $C_6H_5COO^-$
Initial (mol L^{-1})	0.250	—	0	0
Change	$-x$	—	$+x$	$+x$
Equilibrium	$0.250 - x \cong 0.250$	—	x	x

$$K_a \approx \frac{x^2}{0.250} = 6.5 \times 10^{-5}$$

$x = 4.0 \times 10^{-3}$

$$\text{percentage deprotonated} = \frac{4.0 \times 10^{-3}}{0.250} \times 100\% = \underline{1.6\%}$$

(b)

	H_2O	+ NH_2NH_2 $\rightleftharpoons$	$NH_2NH_3^+$ +	OH^-
Initial	—	0.150	0	0
Change	—	$-x$	$+x$	$+x$
Equilibrium	—	$0.150 - x \cong 0.150$	x	x

$$K_b = \frac{[NH_2NH_3^+][OH^-]}{[NH_2NH_2]} \approx \frac{x^2}{0.150} = 1.7 \times 10^{-6}$$

$x = 5.0 \times 10^{-4}$

$$\text{percentage protonated} = \frac{5.0 \times 10^{-4}}{0.150} \times 100\% = \underline{0.33\%}$$

(c)

	$(CH_3)_3N$ +	H_2O $\rightleftharpoons$	$(CH_3)_3NH^+$ +	OH^-
Initial	0.112	—	0	0
Change	$-x$	—	$+x$	$+x$
Equilibrium	$0.112 - x \cong 0.112$	—	x	x

$$K_b = \frac{[(CH_3)_3NH][OH^-]}{[(CH_3)_3N]} \approx \frac{x^2}{0.112} = 6.5 \times 10^{-5}$$

$x = 2.7 \times 10^{-3}$

$$\text{percentage protonated} = \frac{2.7 \times 10^{-3}}{0.112} \times 100\% = \underline{2.4\%}$$

13. Glycine can accept one proton on its nitrogen atom and donate one from its carboxyl group.

    $pK_{a1} = 2.35$, $pK_{a2} = 9.60$

    We follow the procedure of Example 8.4. The three species present are H_2Gly^+, HGly, and Gly^-. The equilibria are:

    $H_2Gly^+(aq) + H_2O(l) \rightleftharpoons H_3O^+(aq) + HGly(aq)$

$$K_{a1} = \frac{[H_3O^+][HGly]}{[H_2Gly^+]} = \frac{H[HGly]}{[H_2Gly^+]} \qquad H = [H_3O^+]$$

$$HGly + H_2O(l) \rightleftharpoons H_3O^+(aq) + Gly^-$$

$$K_{a2} = \frac{H[Gly^-]}{[HGly]}$$

Total concentration = $G = [H_2Gly^+] + [HGly] + [Gly^-]$

$[Gly^-] = K_{a2}[HGly]/H = K_{a2}K_{a1}[H_2Gly^+]/H^2$ because

$[HGly] = K_{a1}[H_2Gly^+]/H$

Then $G = [H_2Gly^+] + K_{a1}[H_2Gly^+]/H + K_{a2}K_{a1}[H_2Gly^+]/H^2$

The fractions are

$$f_1 = f(H_2Gly^+) = \frac{[H_2Gly^+]}{G} = \frac{1}{1 + K_{a1}/H + K_{a2}K_{a1}/H^2}$$

$$= \frac{H^2}{H^2 + K_{a1}H + K_{a2}K_{a1}} = \frac{H^2}{K}$$

where $K = H^2 + K_{a1}H + K_{a2}K_{a1}$

Similarly we find

$$f_2 = f(HGly) = \frac{HK_{a1}}{K}$$

$$f_3 = f(Gly^-) = \frac{K_{a1}K_{a2}}{K}$$

These fractions are plotted in the figure below against pH = –log H. These plots were produced with MathCad 8.

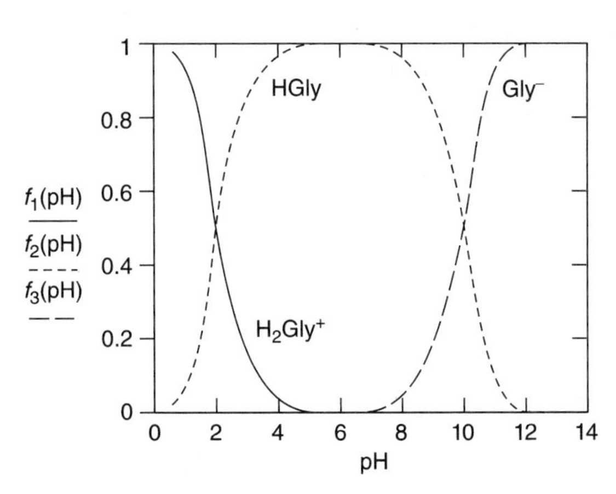

15. For the case of an amphiprotic salt we use the relation

$$\mathrm{pH} = \frac{1}{2}(\mathrm{p}K_{a1} + \mathrm{p}K_{a2})\ [8.10]$$

Note that in this case it is not necessary to specify the concentration of the salt.

For oxalic acid [Table 8.2], pK_{a1} = 1.23 and pK_{a2} = 4.19. Therefore,

$$\mathrm{pH} = \frac{1}{2}(1.23 + 4.19) = \underline{2.71}$$

17. At half neutralization, pH = pK_a, and this is the pH at which the buffering action is best (pH = 8.3).

19. (a)

	$H_2C_2O_4$	+ H_2O ⇌	H_3O^+	+ $HC_2O_4^-$
Initial	0.15	—	0	0
Change	$-x$	—	$+x$	$+x$
Equilibrium	$0.15 - x$	—	x	x

$K_{a1} = 5.9 \times 10^{-2}$, $K_{a2} = 6.5 \times 10^{-5}$
The second ionization can be ignored in the calculation of $[H_3O^+]$, but not in the calculation of $[C_2O_4^-]$.

$$K_{a1} = 5.9 \times 10^{-2} = \frac{x^2}{0.15 - x}$$

$x^2 + (5.9 \times 10^{-2})x - (8.85 \times 10^{-3}) = 0$

$x = \underline{0.069\ \mathrm{mol\ L^{-1}}} = [H_3O^+]$

$[OH^-] = \underline{1.4 \times 10^{-13}\ \mathrm{mol\ L^{-1}}}$

$[H_2C_2O_4] = 0.15 - 0.069 = \underline{0.08\ \mathrm{mol\ L^{-1}}}$

	$HC_2O_4^-$	+ H_2O ⇌	H_3O^+	+ $C_2O_4^{2-}$
Initial	0.069	—	0.069	0
Change	$-x$	—	$+x$	$+x$
Equilibrium	$0.069 - x$	—	$0.069 + x$	x

$$K_{a2} = 6.5 \times 10^{-5} = \frac{(0.069 + x)x}{(0.069 - x)} \approx x \text{ [because } x \text{ is small]}$$

$x = [C_2O_4^{2-}] = 6.5 \times 10^{-5}$ mol L^{-1} and

$[HC_2O_4^-] = 0.069 - x = 0.069 - 0.000065 = \underline{0.069 \text{ mol L}^{-1}}$

(b) Similar to part a, K_{a2} can be ignored in the first part of the calculation and

$$K_{a1} = 1.3 \times 10^{-7} = \frac{x^2}{0.065 - x} \approx \frac{x^2}{0.065}$$

$[H_2S] = \underline{0.065 \text{ mol L}^{-1}}$

$x = [H_3O^+] = [HS^-] = \underline{9.2 \times 10^{-5} \text{ mol L}^{-1}}$

$$[OH^-] = \frac{1.0 \times 10^{-14}}{9.2 \times 10^{-5}} = \underline{1.1 \times 10^{-10} \text{ mol L}^{-1}}$$

For the second ionization, $K_{a2} = 7.1 \times 10^{-15} = x = [S^{2-}]$

$[S^{2-}] = \underline{7.1 \times 10^{-15} \text{ mol L}^{-1}}$

21. $$K_a = \frac{[H_3O^+][CH_3CO_2^-]}{[CH_3COOH]}$$

$$pH = pK_a + \log \frac{[CH_3CO_2^-]}{[CH_3COOH]}$$

(a) $$pH = pK_a + \log \frac{[0.10]}{[0.10]} = pK_a = \underline{4.74}$$

(b) 3.3 mmol NaOH = 3.3×10^{-3} mol OH^- [strong base] produces 3.3×10^{-3} mol CH_3COO^- from CH_3COOH

Initially $n\,(CH_3COOH) = n\,(CH_3COO^-) = 0.10$ mol L$^{-1} \times 0.100$ L $= 1.0 \times 10^{-2}$ mol

After adding NaOH

$$[CH_3COOH] = \frac{1.0 \times 10^{-2} - 3.3 \times 10^{-3}}{0.10 \text{ L}} = 6.7 \times 10^{-2} \text{ mol L}^{-1}$$

$$[CH_3COO^-] = \frac{1.0 \times 10^{-2} + 3.3 \times 10^{-3}}{0.10 \text{ L}} = 0.13 \text{ mol L}^{-1}$$

$$pH = 4.74 + \log \frac{0.13}{0.067} = \underline{5.03}$$

Change in pH = $\underline{0.29}$

(c) 6.0 mmol HNO_3 = 6.0×10^{-3} mmol H_3O^+ [strong acid] produces 6.0×10^{-3} mol CH_3COOH from CH_3COO^-

after adding HNO_3 [see (b) above]

$$[CH_3COOH] = \frac{(1.0 \times 10^{-2}) + (6.0 \times 10^{-3})\text{ mol}}{0.100\text{ L}} = 0.16\text{ mol L}^{-1}$$

$$[CH_3COO^-] = \frac{(1.0 \times 10^{-2}) - (6.0 \times 10^{-3})\text{ mol}}{0.100\text{ L}} = 4.0 \times 10^{-2}\text{ mol L}^{-1}$$

$$\text{pH} = 4.74 + \log\frac{4.0 \times 10^{-2}}{0.16} = 4.74 - 0.60 = \underline{4.14}$$

change in pH = $\underline{-0.60}$

23. At the halfway point, pH = pK_a = $\underline{4.66}$

and $K_a = \underline{2.19 \times 10^{-5}}$

$$K_a = 2.19 \times 10^{-5} = \frac{x^2}{(0.015 - x)} \approx \frac{x^2}{(0.015)}$$

$x = 5.7 \times 10^{-4}$ M = $[H_3O^+]$

pH = $\underline{3.24}$

25. At the stoichiometric point the solution will consist of the lactate ion, which is a weak base, and Na^+ ions. To calculate the pH we first need to calculate the total volume of solution at the stoichiometric point.

amount (moles) lactate ion = 0.02500 L × 0.100 mol L^{-1}

= 2.50×10^{-3} mol

volume of base added = $(2.500 \times 10^{-3}$ mol)/(0.175 mol L^{-1})

= 0.0143 L

total volume = 0.02500 L + 0.0143 L = 0.0393 L

[lactate ion] = $(2.50 \times 10^{-3}$ mol)/(0.0393 L) = 0.0636 mol L^{-1}

$$K_b = 1.2 \times 10^{-11} = \frac{[\text{lactic acid}]\,[OH^-]}{[\text{lactate ion}]} = \frac{x^2}{0.0636 - x} \approx \frac{x^2}{0.0636}$$

$x = [OH^-] = 8.7 \times 10^{-7}$ M

pOH = $-\log(8.7 \times 10^{-7}) = 6.06$

pH = 14.00 − 6.06 = $\underline{7.94}$

27. Choose a buffer system in which the conjugate acid has a pK_a close to the desired pH. Therefore,

(a) H_3PO_4 and NaH_2PO_4

(b) NaH_2PO_4 and Na_2HPO_4 or $NaHSO_3$ and Na_2SO_3

29. (a) $K_s = [Ba^{2+}][SO_4^{2-}] = S^2 = 1.1 \times 10^{-10}$

$S = \underline{1.0 \times 10^{-5}\ \text{mol L}^{-1}}$

(b) $K_s = [Ag^+]^2[CrO_4^{2-}] = (2S)^2(s) = 4S^3 = 9.0 \times 10^{-12}$

$S = \underline{1.3 \times 10^{-4}\ \text{mol L}^{-1}}$

(c) $K_s = [Fe^{3+}][OH^-]^3 = (S)(3S)^3 = 27S^4 = 2.0 \times 10^{-39}$

$S = \underline{9.3 \times 10^{-11}\ \text{mol L}^{-1}}$

(d) $K_s = [Hg_2^{2+}][SO_4^{2-}] = S^2 = 6.8 \times 10^{-7}$

$S = 8.2 \times 10^{-4}$

31. $HgCl_2(s) \rightleftharpoons Hg^{2+}(aq) + 2Cl^-(aq) \quad K = [Hg^{2+}][Cl^-]^2$

$[Cl^-] = 2 \times [Hg^{2+}]$; therefore $K = 4[Hg^{2+}]^3$

and the solubility of the salt is

$S = [Hg^{2+}] = (\frac{1}{4}K)^{1/3}\ \text{mol L}^{-1}$

From $\Delta_r G^\ominus = \Delta_f G^\ominus(Hg^{2+}) + 2\Delta_f G^\ominus(Cl^-) - \Delta_f G^\ominus(HgCl_2)$

$= +164.40 + 2 \times (-131.23) - (-178.6)\ \text{kJ mol}^{-1}$

$= 80.54\ \text{kJ mol}^{-1}$

$$\ln K = \frac{-\Delta_r G^\ominus}{RT} = \frac{-80.54 \times 10^3\ \text{J mol}^{-1}}{8.315\ \text{J K}^{-1}\ \text{mol}^{-1} \times 298.15\ K} = -32.49$$

Therefore $K = 7.758 \times 10^{-15}$

and $S = \underline{1.25 \times 10^{-5}\ \text{mol L}^{-1}}$

Electrochemistry

1. Yes, NADH is oxidized (it loses electrons); pyruvate $CH_3COCO_2^-$, is reduced (it gains electrons).

3. (3) $CH_3CH_2OH + NAD^+ \rightarrow CH_3CHO + NADH + H^+$

   (1) $NAD^+ + 2e^- + H^+ \rightarrow NADH$

   (2) $CH_3CHO + 2e^- + 2H^+ \rightarrow CH_3CH_2OH$

   Reaction (3) is obtained as (1) – (2). The reaction quotients for each of these reactions are:

$$Q_{(1)} = \frac{[NADH]}{[NAD^+][H^+]}$$

$$Q_{(2)} = \frac{[CH_3CH_2OH]}{[CH_3CHO][H^+]^2}$$

$$Q_{(3)} = \frac{[CH_3CHO][NADH][H^+]}{[CH_3CH_2OH][NAD^+]}$$

   It is seen that $Q_{(3)} = \dfrac{Q_{(1)}}{Q_{(2)}}$

5. (1) $NADP^+(aq) + H^+(aq) + 2e^- \rightarrow NADPH(aq)$

   Subtraction of this half reaction from the overall reaction yields

   $2fd_{red}(aq) + H^+(aq) - 2e^- \rightarrow 2fd_{ox}(aq)$

   which may be rearranged to

   (2) $2fd_{ox}(aq) + 2e^- \rightarrow 2fd_{red}(aq) + H^+(aq)$

   Then the overall reaction is obtained as (1) – (2), with 2 electrons transferred.

7. The Nernst equation applies to half-cells as well as cells.

$$E_i = E^{\ominus} - \frac{RT}{\nu F} \ln Q_i \ [9.12]$$

$$= E^{\ominus} - \frac{25.7\ \text{mV}}{\nu} \ln Q_i$$

$$E_f = E^{\ominus} - \frac{25.7\ \text{mV}}{\nu} \ln Q_f$$

The half-reaction for the electrode is

$2H^+(aq) + 2e^- \rightarrow H_2(g)\ E^{\ominus} = 0$

$$Q = \frac{p_{H_2}}{[H^+]^2} = \frac{1.45}{[H^+]^2}$$

$$\Delta E = E_f - E_i = -\frac{25.7\ \text{mV}}{2} \ln\left(\frac{Q_f}{Q_i}\right)$$

$$= -25.7\ \text{mV} \ln \frac{[H^+]_i}{[H^+]_f}$$

$$= 25.7\ \text{mV} \ln \frac{[H^+]_f}{[H^+]_i}$$

$$= 25.7\ \text{mV} \ln \left(\frac{25.0}{5.0}\right)$$

$$= \underline{41\ \text{mV}}$$

9. R: $Cl_2(g) + 2e^- \rightarrow 2Cl^-(aq)$ $\quad E^{\ominus} = +1.36V$

L: $Mn^{2+}(aq) + 2e^- \rightarrow Mn(s)$ $\quad E^{\ominus} = ?$

The cell corresponding to these half reactions is

$Mn|MnCl_2(aq)|Cl_2(g)|Pt$ $\quad E^{\ominus} = 1.36V - E^{\ominus}(Mn^{2+}, Mn)$

Hence, $E^{\ominus}(Mn^{2+}, Mn) = 1.36V - 2.54V = \underline{-1.18V}$

11. (a) $E = E^{\ominus} - \frac{RT}{F} \ln \frac{b_L}{b_R}$

(b) $E = E^{\ominus} - \frac{RT}{2F} \ln \frac{p_R}{p_L}$

In the following Nernst equations involving ions in aqueous solution we have replaced activities with molar concentrations

(c) $E = E^{\ominus} - \dfrac{RT}{2F} \ln \dfrac{[Mn^{2+}][Fe(CN)_6^{3-}]^2}{[H^+]^4[Fe(CN)_6^{4-}]^2}$

(d) $E = E^{\ominus} - \dfrac{RT}{2F} \ln \dfrac{p_{Cl_2}[Br^-]^2}{[Cl^-]^2}$

(e) $E = E^{\ominus} - \dfrac{RT}{2F} \ln \dfrac{[Sn^{2+}][Fe^{3+}]^2}{[Sn^{4+}][Fe^{2+}]^2}$

(f) $E = E^{\ominus} - \dfrac{RT}{2F} \ln \dfrac{[Fe^{2+}][Mn^{2+}]}{[H^+]^4}$

13. (a) This reaction is analyzed in Exercise 9.3. $\underline{\nu = 2}$

A possible cell arrangement is

$Pt | CH_3CH_2OH(aq), CH_3CHO(aq), H^+(aq) || H^+(aq), NAD^+(aq), NADH(aq) | Pt$

(b) R: $Mg^{2+}(aq) + 2e^- \rightarrow Mg(s)$

L: $MgATP^{2-} + 2e^- \rightarrow Mg(s) + ATP^{4-}$

$Mg(s) | ATP^{4-}(aq), MgATP^{2-}(aq) || Mg^{2+}(aq) | Mg(s)$ $\underline{\nu = 2}$

(c) R: $CH_3COCO_2^-(aq) + 2e^- + 2H^+(aq) \rightarrow CH_3CH(OH)CO_2^-(aq)$

L: Cyt-c(ox, aq) + $e^- \rightarrow$ Cyt-c(red, aq)

The overall reaction is obtained from R – 2 × L. $\underline{\nu = 2}$

Pt | Cyt-c(red, aq), Cty-c (ox, aq) || $H^+(aq), CH_3CH(OH)CO_2^-(aq), CH_3COCO_2^-(aq) | Pt$

15. See the solution to Exercise 9.12

(a) $E^{\ominus}_{cell} = -0.36\ V - (-0.44\ V) = \underline{+0.08\ V}$

(b) $E^{\ominus}_{cell} = +0.27\ V - 0\ V = \underline{+0.27\ V}$

(c) $E^{\ominus}_{cell} = +1.23\ V - 0\ V = \underline{+1.23\ V}$

(d) $E^{\ominus}_{cell} = +0.695\ V - 0\ V = \underline{+0.695\ V}$

(e) $E^{\ominus}_{cell} = +0.54\ V - 0\ V = \underline{+0.54\ V}$

(f) $E^{\ominus}_{cell} = +0.52\ V - 0.15\ V = \underline{+0.37\ V}$

17. (a) E decreases, $E = E^{\ominus} - \dfrac{RT}{F} \ln\left(\dfrac{[Ag^+]_L}{[Ag^+]_R}\right)$

(b) E increases, $E = E^{\ominus} - \dfrac{RT}{2F} \ln\left(\dfrac{p_R}{p_L}\right)$

(c) E increases, $E = E^{\ominus} - \dfrac{RT}{2F} \ln\left(\dfrac{[Mn^{2+}][Fe(CN)_6^{3-}]^2}{[H^+]^4[Fe(CN)_6^{4}]^2}\right)$

(d) E increases, $E = E^{\ominus} - \dfrac{RT}{2F} \ln\left(\dfrac{[Br^-]^2 p_{Cl_3}}{[Cl^-]^2}\right)$

(e) E decreases, $E = E^{\ominus} - \dfrac{RT}{2F} \ln\left(\dfrac{[Sn^{2+}][Fe^{3+}]^2}{[Sn^{4+}][Fe^{2+}]^2}\right)$

(f) E increases, $E = E^{\ominus} - \dfrac{RT}{2F} \ln\left(\dfrac{[Fe^{2+}][Mn^{2+}]}{[H^+]^4}\right)$

19. (a) The Nernst equation is

$$E = E^{\ominus} - \frac{RT}{2F} \ln \frac{[CH_3CHO][NADH][H^+]}{[CH_3CH_2OH][NAD^+]}$$

Increasing pH implies decreasing $[H^+]$; therefore the cell potential <u>decreases</u>.

(b) The Nernst equation is

$$E = E^{\ominus} - \frac{RT}{2F} \ln \frac{[MgATP^{2-}]}{[ATP^{4-}][Mg^{2+}]}$$

Increasing $[Mg^{2+}]$ from $MgSO_4$ <u>increases</u> the cell potential.

(c) The Nernst equation is

$$E = E^{\ominus} - \frac{RT}{2F} \ln \frac{[\text{Cyt-c(ox)}][CH_3CH(OH)CO_2^-]}{[\text{Cyt-c(red)}][CH_3COCO_2^-][H^+]^2}$$

Increasing $[CH_3CH(OH)CO_2^-]$ <u>decreases</u> the cell potential.

21. Each section has to be broken down into the two half-reactions from which it is formed. The standard potential for the cell is then calculated from the standard electrode potentials in Appendix 2. The standard Gibbs energies of the reaction are then calculated from $\Delta_r G^{\ominus} = -\nu F E^{\ominus}$.

(a) (1) $Ca^{2+}(aq) + 2e^- \rightarrow Ca(s)$ $E_1^{\ominus} = -2.87$ V

(2) $2H_2O(l) + 2e^- \rightarrow H_2(g) + 2OH^-(aq)$ $E_2^{\ominus} = -0.83$ V

The overall reaction is then obtained from (2) – (1) and the $E^{\ominus}$ value for the cell reaction is calculated in the same manner, that is,

$E^{\ominus} = E_2^{\ominus} - E_1^{\ominus} = -0.83\text{ V} - (-2.87\text{ V}) = +2.04\text{ V}$

$\Delta_r G^{\ominus} = -\nu F E^{\ominus}$

$= -2 \times 96.485\text{ kC mol}^{-1} \times 2.04\text{ V} = \underline{-394\text{ kJ mol}^{-1}}$

(b) same as above

The standard Gibbs energies of reaction for reactions (c), (d), (e), and (f) are obtained by a procedure similar to that described for part (a). The results are:

(c) $E^{\ominus} = -0.39$ V

Therefore $\Delta_r G^{\ominus} = -\nu F E^{\ominus}$

$= -2 \times 96.485\text{ kC mol}^{-1} \times (-0.39\text{ V}) = \underline{+75\text{ kJ mol}^{-1}}$

(d) $E^{\ominus} = +1.51$ V

Therefore $\Delta_r G^{\ominus} = -\nu F E^{\ominus}$

$= -2 \times 96.485\text{ kC mol}^{-1} \times 1.51\text{ V} = \underline{-291\text{ kJ mol}^{-1}}$

(e) same as above

(f) $E^{\ominus} = -2.58$ V

Therefore $\Delta_r G^{\ominus} = -\nu F E^{\ominus}$

$= -2 \times 96.485\text{ kC mol}^{-1} \times (-2.58\text{ V}) = \underline{-498\text{ kJ mol}^{-1}}$

23. (a) $E^{\ominus} = \dfrac{-\Delta_r G^{\ominus}}{\nu F} = \dfrac{+62.5\text{ kJ mol}^{-1}}{2 \times 96.485\text{ kC mol}^{-1}} = \underline{+0.324\text{ V}}$

(b) $E^{\ominus} = E^{\ominus}(Fe^{3+}, Fe^{2+}) - E^{\ominus}(Ag_2CrO_4, Ag, CrO_4^{2-})$

Therefore,

$E^{\ominus}(Ag_2CrO_4, Ag, CrO_4^{2-}) = E^{\ominus}(Fe^{3+}, Fe^{2+}) - E^{\ominus}$

$= +0.77 - 0.324\text{ V} = \underline{+0.45\text{ V}}$

25. $2Ag(s) + Cu^{2+}(aq) \rightarrow 2Ag^+(aq) + Cu(s)$

$E = E^{\ominus}(Cu^{2+}, Cu) - E^{\ominus}(Ag^+, Ag) = +0.34 - 0.80\text{ V} = \underline{-0.46\text{ V}}$

$\Delta_r G^{\ominus} = 2\Delta_f G^{\ominus}(Ag^+, aq) - \Delta_f G^{\ominus}(Cu^{2+}, aq)$

$= [(2 \times 77.1) - (64.49)]\text{ kJ mol}^{-1} = \underline{+89.7\text{ kJ mol}^{-1}}$

Alternatively, $\Delta_r G^{\oplus} = \nu F E^{\oplus}$

$= -(2)(96.485\text{ kC mol}^{-1})\ (-0.46\text{ V})$

$= +88.8\text{ kJ mol}^{-1}$

$$\Delta_r H^{\ominus} = 2\,\Delta_f H^{\ominus}\,(Ag^+, aq) - \Delta_f H^{\ominus}\,(Cu^{2+}, aq)$$

$$= [(2 \times 105.58) - (64.77)]\ \text{kJ mol}^{-1} = \underline{146\ \text{kJ mol}^{-1}}$$

$$\Delta_r S^{\ominus} = \frac{\Delta_r G^{\ominus} - \Delta_r H^{\ominus}}{T} \quad [\Delta_r G = \Delta_r H - T\Delta_r S]$$

$$\Delta_r S^{\ominus} = 188\ \text{J mol}^{-1}\ \text{K}^{-1}$$

Therefore $\Delta_r G^{\ominus}\,(308\ \text{K}) \approx [146 - 308 \times 0.188]\ \text{kJ mol}^{-1}$

$$\approx \underline{+88\ \text{kJ mol}^{-1}}$$

27. We follow the procedure of Example 9.6.

The half-cell reactions are:

R: $O_2(g) + 4H^+(aq) + 4e^- \rightarrow 2H_2O(l)$

L: cystine (aq) + $2H^+(aq) + 2e^- \rightarrow$ 2 cysteine (aq)

For the right electrode

$$E = E^{\ominus} - \frac{RT}{4F} \ln\left(\frac{a^2_{H_2O}}{p_{O_2} a^4_{H^+}}\right)$$

We set a_{H_2O} and p_{O_2} equal to their standard values

$$E = E^{\ominus} - \frac{RT}{4F} \ln\left(\frac{1}{a^4_{H^+}}\right) = E^{\ominus} + \frac{RT}{F} \ln a_{H^+}$$

$$\ln a_{H^+} = -\ln 10\ \text{pH} = -2.303\ \text{pH}$$

$$E = E^{\ominus} - \frac{2.303\,RT}{F} \times \text{pH}$$

$$E = E^{\ominus} - \frac{2.303 \times 8.315\ \text{J K}^{-1}\ \text{mol}^{-1} \times 310\ \text{K}}{96485\ \text{C mol}} \times \text{pH}$$

$$E^{\oplus} = 1.23\ \text{V} - 0.0615\ \text{V} \times 7 = \underline{+0.80\ \text{V}}$$

For the left electrode after following a similar procedure

$$E = E^{\oplus} - 0.0615\ \text{V} \times \text{pH}$$

$$E^{\oplus} = -0.34\ \text{V} - 0.0615\ \text{V} \times 7 = \underline{-0.77\ \text{V}}$$

For the system as a whole

$$E_{cell} = E_R - E_L = 0.80\ \text{V} - (-0.77\ \text{V}) = \underline{+1.57\ \text{V}}$$

29. The couple is

$CH_3COCOOH + 2H^+ + 2e^- \rightarrow CH_3CH(OH)COOH$

The Nernst equation for the couple is

$$E = E^{\ominus} - \frac{RT}{2F} \ln \frac{a_{HL}}{a_{HP}a^2_{H+}}$$

We assume $a_{HL} = a_{HP} = 1$, then

$$E = E^{\ominus} + \frac{RT}{F} \ln a_{H+} = E^{\ominus} - \frac{2.303\ RT}{F} \times pH$$

$$= E^{\ominus} - 0.0615\ V \times pH\ [\text{for } T = 310\ K]$$

$$E = -0.19\ V = E^{\ominus} - 0.0615\ V \times 7$$

$$E^{\ominus} = -0.19\ V + 0.43\ V = \underline{+0.24\ V}$$

31. $Cu_3(PO_4)_2(s) \rightleftharpoons 3Cu^{2+}(aq) + 2PO_4^{3-}(aq)$

(a) $K_s = [Cu^{2+}]^3[PO_4^{3-}]^2$

The molar solubility of the salt is S, $[Cu^{2+}] = 3S$, $[PO_4^{3-}] = 2S$

Therefore

$K_s = 1.3 \times 10^{-37} = (3S)^3(2S)^2 = 108S^5$

$S = 1.6 \times 10^{-8}\ mol\ L^{-1}$

(b) The cell reaction is

R: $Cu^{2+}(aq) + 2e^- \rightarrow Cu(s)$	+0.34 V
L: $2H^+(aq) + 2e^- \rightarrow H_2(g)$	0
Overall: $Cu^{2+}(aq) + H_2(g) \rightarrow Cu(s) + 2H^+(aq)$	+0.34 V

Using the Nernst equation

$$E = E^{\ominus} \frac{RT}{\nu F} \ln Q$$

$$= 0.34\ V - \frac{25.693 \times 10^{-3}\ V}{2} \ln \frac{a^2_{H+}}{a_{Cu2+}}$$

Because $a_{Cu2+} \approx [Cu^{2+}] = 3S$, the following substitution can be made:

$$E = 0.34\ V - \frac{25.693 \times 10^{-3}\ V}{2} \ln \frac{1}{3 \times (1.6 \times 10^{-8})}$$

$$= 0.34\ V - 0.22\ V = \underline{+0.12\ V}$$

33. $S(\text{AgCl}) = b(\text{Ag}^+)$

$\text{AgCl(s)} \rightleftharpoons \text{Ag}^+\text{(aq)} + \text{Cl}^-\text{(aq)} \qquad K_s \approx b(\text{Ag}^+)b(\text{Cl}^-)/b^{\ominus 2}$

Because $b(\text{Ag}^+) = b(\text{Cl}^-)$

$K_s \approx b(\text{Ag}^+)^2/\, b^{\ominus 2} = S^2/\, b^{\ominus 2} = (1.34 \times 10^{-5})^2$

$\approx \underline{1.80 \times 10^{-10}}$

$S(\text{BaSO}_4) = b(\text{Ba}^+)$

$\text{BaSO}_4\text{(s)} \rightleftharpoons \text{Ba}^{2+}\text{(aq)} + \text{SO}_4^{2-}\text{(aq)}$

As above, $K_s \approx S^2/\, b^{\ominus 2} = (9.51 \times 10^{-4})^2 = \underline{9.04 \times 10^{-7}}$

$\text{AgCl(s)} \rightleftharpoons \text{Ag}^+\text{(aq)} + \text{Cl}^-\text{(aq)}$

This reaction can be constructed from the following half-cell reactions.

R: $\text{AgCl(s)} + \text{e}^- \rightarrow \text{Ag(s)} + \text{Cl}^-\text{(aq)} \qquad E^{\ominus}_{\text{AgCl, Ag, Cl}^-} = ?$

L: $\text{Ag}^+\text{(aq)} + \text{e}^- \rightarrow \text{Ag(s)} \qquad E^{\ominus}_{\text{Ag}^+,\text{Ag}} = 0.80\ \text{V}$

R–L: $E^{\ominus}_{\text{cell}} = (E^{\ominus}_{\text{AgCl, Ag, Cl}^-} - 0.80\ \text{V})$

$\Delta_r G^{\ominus} = -RT \ln K_s = -\nu F E^{\ominus}_{\text{cell}}$

$$\ln K_s = \frac{\nu F E^{\ominus}_{\text{cell}}}{RT} = \frac{(E^{\ominus}_{\text{AgCl, Ag, Cl}^-} - 0.80\ \text{V})}{0.0257\ \text{V}}$$

$E^{\ominus}_{\text{AgCl, Ag, Cl}^-} = 0.0257\ \text{V} \times \ln K_s + 0.80\ \text{V}$

$= 0.0257\ \text{V} \times \ln(1.80 \times 10^{-10}) + 0.80\ \text{V}$

$= \underline{+0.22\ \text{V}}$

$\text{BaSO}_4\text{(s)} \rightleftharpoons \text{Ba}^{2+}\text{(aq)} + \text{SO}_4^{2-}\text{(aq)}$

This reaction can be constructed from

R: $\text{BaSO}_4\text{(s)} + 2\text{e}^- \rightarrow \text{Ba(s)} + \text{SO}_4^{2-}\text{(aq)} \qquad E^{\ominus}_{\text{BaSO}_4} = ?$

L: $\text{Ba}^{2+}\text{(aq)} + 2\text{e}^- \rightarrow \text{Ba(s)} \qquad E^{\ominus}_{\text{Ba}^+,\text{Ba}} = -2.91\ \text{V}$

R–L: $E^{\ominus}_{\text{cell}} = (E^{\ominus}_{\text{BaSO}_4} + 2.91\ \text{V})$

$\Delta_r G^{\ominus} = -RT \ln K_s = -\nu F E^{\ominus}_{\text{cell}}$

$$\ln K_s = \frac{\nu F E^{\ominus}_{\text{cell}}}{RT} = 2 \times \frac{(E^{\ominus}_{\text{BaSO}_4} + 2.91\ \text{V})}{0.0257\ \text{V}}$$

$E^{\ominus}_{\text{BaSO}_4} = (0.0257\ \text{V} \times \ln K_s - 2.91\ \text{V})/2$

$= [0.0257\ \text{V} \times \ln(9.04 \times 10^{-7}) - 2.91\ \text{V}]/2$

$= \underline{-1.63\ \text{V}}$

35. R: $2AgCl(s) + 2e^- \rightarrow 2Ag(s) + 2Cl^-(aq)$ $+0.22$ V

L: $2H^+(aq) + 2e^- \rightarrow H_2(g)$ 0

Overall: $2AgCl(s) + H_2(g) \rightarrow 2Ag(s) + 2Cl^-(aq) + 2H^+(aq)$

$Q = a_{H+}^2 a_{Cl-}^2$ $\nu = 2$

$= a_{H+}^4$ $a_{H+} = a_{Cl-}$

Therefore from the Nernst equation,

$$E = E^{\ominus} - \frac{RT}{2F} \ln a_{H+}^4 = E^{\ominus} - \frac{2RT}{F} \ln a_{H+} = 0.312\ \text{V}$$

$$= E^{\ominus} + 2.303 \frac{2RT}{F} \text{pH}$$

Rearranging and substituting in:

$$\text{pH} = \frac{F}{2 \times 2.303\ RT} \times (E - E^{\ominus}) = \frac{0.312\ \text{V} - 0.22\ \text{V}}{0.1183\ \text{V}}$$

$= \underline{0.78}$

37. R: $Ag^+(aq) + e^- \rightarrow Ag(s)$ $E^{\ominus} = +0.80$ V

L: $AgI(s) + e^- \rightarrow Ag(s) + I^-(aq)$ $E^{\ominus} = \underline{-0.15\ \text{V}}$

Overall: $Ag^+(aq) + I^-(aq) \rightarrow AgI(s)$ $\nu = 1$ $E^{\ominus} = +0.95\ \text{V} \approx 0.9509$ V

$$\ln K = \frac{0.9509\ \text{V}}{25.693\ \text{mV}} = 37.010,\ K = 1.184 \times 10^{16}$$

However, $K_s = K^{-1}$ because the solubility equilibrium is written as the reverse of the cell reaction. Therefore, $K_s = \underline{8.5 \times 10^{17}}$.

The solubility is obtained from $[Ag^+] = [I^-]$ and $S = [Ag^+]$

$K_s = [Ag^+]^2$, so that

$S = (K_s)^{1/2}\ \text{mol L}^{-1} = (8.45 \times 10^{-17})^{1/2}\ \text{mol L}^{-1}$

$= \underline{9.19 \times 10^{-9}\ \text{mol L}^{-1}}$

39. (a) $H_2(g) + 1/2\ O_2(g) \rightarrow H_2O(l)$

$\Delta_r G^{\ominus} = \Delta_f G^{\ominus}(H_2O, l) = -237.13\ \text{kJ mol}^{-1}$

$$E^{\ominus} = -\frac{\Delta_r G^{\ominus}}{\nu F} = \frac{+237.13\ \text{kJ mol}^{-1}}{2 \times (96.485\ \text{kC mol}^{-1})} = \underline{+1.23\ \text{V}}$$

(b) $C_6H_6(g) + 7\frac{1}{2}\ O_2(g) \rightarrow 6CO_2(g) + 3H_2O(l)$

$$\Delta_r G^{\ominus} = 6\ \Delta_f G^{\ominus}(CO_2, g) + 3\ \Delta_f G^{\ominus}(H_2O, l) - \Delta_f G^{\ominus}(C_6H_6, g)$$
$$= [6 \times (-394.36) + 3 \times (-237.13) - (+129.7)]\text{kJ mol}^{-1}$$
$$= -3207\ \text{kJ mol}^{-1}$$

To find the number of electrons transferred, note that the cathode half-reaction is the reduction of oxygen to produce $6CO_2 + 3H_2O$:

$$7\frac{1}{2}\ O_2(g) + 30\ e^- \rightarrow 15O_2^-$$

Therefore,

$$E^{\ominus} = -\frac{\Delta_r G^{\ominus}}{\nu F} = \frac{+3207\ \text{kJ mol}^{-1}}{30 \times 96.485\ \text{kC mol}^{-1}} = \underline{+1.108\ \text{V}}$$

41. 41. $MnO_4^- + 8H^+ + 5e^- \rightarrow Mn^{2+} + 4H_2O$ $\qquad E^{\ominus} = 1.51\ \text{V}$

(a) Using the Nernst equation to determine the reduction potential for pH 7.00, keeping $a_{MnO_4^-}$ and $a_{Mn^{2+}} = 1.00$.

$$E = E^{\ominus} - \frac{25.7\ \text{mV}}{\nu}\ \text{In}\ Q$$

$$\text{In}\ Q = \ln \frac{1}{a_{H^+}^8} = 8 \times 2.303 \times \text{pH}$$

$$E = 1.51\ \text{V} - \frac{25.7\ \text{mV} \times 8 \times 2.303 \times 6.00}{5} = \underline{+0.94\ \text{V}}$$

(b) In general, $\underline{E = 1.51 - 0.0947\ \text{pH}}$

For the left electrode after following a similar procedure

$$E = E^{\ominus} - 0.0615\ \text{V} \times \text{pH}$$
$$= -0.34\ \text{V} - 0.0615\ \text{V} \times 7 = \underline{-0.77\ \text{V}}$$

For the system as a whole

$$E_{cell} = E_R - E_L = 0.80\ \text{V} - (-0.77\ \text{V}) = \underline{+1.57\ \text{V}}$$

The Rates of Reaction

10

1. Because the rate of formation of C is known, the reaction stoichiometry can be used to determine the rates of consumption and formation of the other participants in the reaction.

$$\text{rate of consumption of A} = \frac{2}{3} \times \text{rate of formation of C}$$
$$= \underline{1.5 \text{ mol L}^{-1} \text{ s}^{-1}}$$

$$\text{rate of consumption of B} = \frac{1}{3} \times \text{rate of formation of C}$$
$$= \underline{0.73 \text{ mol L}^{-1} \text{ s}^{-1}}$$

$$\text{rate of formation of D} = \frac{2}{3} \times \text{rate of formation of C}$$
$$= \underline{1.5 \text{ mol L}^{-1} \text{ s}^{-1}}$$

3. (a) concentrations in (molecules m^{-3}) = $N\ \text{m}^{-3}$

(i) second-order

rate = $N\ \text{m}^{-3}\ \text{s}^{-1} = [k] \times (N\ \text{m}^{-3})^2$

where $[k]$ = units of k and N = number of molecules

$$\text{Then, } [k] = \frac{N\ \text{m}^{-3}\ \text{s}^{-1}}{(N\ \text{m}^{-3})^2} = N^{-1}\ \text{m}^3\ \text{s}^{-1}$$

Because N is unitless, $[k] = \underline{\text{m}^3\ \text{s}^{-1}}$,

though loosely we may say $\text{molecules}^{-1}\ \text{m}^3\ \text{s}^{-1}$

$$\text{(ii) third-order } [k] = \frac{N\ \text{m}^{-3}\ \text{s}^{-1}}{(N\ \text{m}^{-3})^3} = N^{-2}\ \text{m}^6\ \text{s}^{-1}$$

$$\text{or } [k] = \underline{\text{m}^6\ \text{s}^{-1}}$$

(b) pressures in kilopascals

(i) second-order

rate = kPa s^{-1} = $[k] \times (\text{kPa})^2$

$$[k] = \frac{\text{kPa s}^{-1}}{(\text{kPa})^2} = \underline{\text{kPa}^{-1}\ \text{s}^{-1}}$$

(ii) third-order

$$[k] = \frac{\text{kPa s}^{-1}}{(\text{kPa})^3} = \underline{\text{kPa}^{-2}\ \text{s}^{-1}}$$

5. (a) We convert the concentration of the complex to mol L^{-1} and fit the data first to

(1) $\log r_0 = \log k' + a \log [\text{complex}]$
where $k' = k[\text{Y}]_0^b$, then to
(2) $\log k' = \log k + b \log [\text{Y}]_0$

log [complex]	−2.096	−2.035	−1.917
$\log r_0$(a)	2.097	2.158	2.279
$\log r_0$(b)	2.806	2.863	2.982
$\log [\text{Y}]_0$	(a) −2.569	(b) −2.215	

From the fit of the data to equation (1) we obtain

$a = 1.018 \approx 1$ for data (a), $\log k' = 4.230$

$a = 0.987 \approx 1$ for data (b), $\log k' = 4.873$

The order of the reaction with respect to [complex] is seen to be $\underline{1}$.

From the fit of the data to equation (2) we obtain $b = 1.816$

This value is closest to 2, so we suggest that the order with respect to [Y] is $\underline{2}$, but this is not a very satisfying result.

(b) The value of $\log k$ obtained from the fit is

$\log k = 8.896$

and then $k = 7.87 \times 10^8\ \text{L}^2\ \text{mol}^{-2}\ \text{s}^{-1}$. Solving for k from equation (2) after setting $b = 2$

$$\log k = \log k' - 2 \log [\text{Y}]_0$$
$$= 4.230 - 2\,(-2.569)\ \text{[Data from (a)]}$$
$$= 9.368$$

$k = \underline{2.33 \times 10^9\ \text{L}^2\ \text{mol}^{-2}\ \text{s}^{-1}}$

Using data from part (b), the result is $k = 2.09 \times 10^9\ \text{L}^2\ \text{mol}^{-2}\ \text{s}^{-1}$. This can be considered satisfactory agreement in view of the fact that b was found to be 1.816 rather than 2.

7. Use $\ln \frac{[A]_0}{[A]} = kt$ [10.7a]

Solve for k

$$k = \frac{\ln\left(\frac{220}{56.0}\right)}{1.22 \times 10^4\ \text{s}} = \underline{1.12 \times 10^{-4}\ \text{s}^{-1}}$$

9. $\frac{1}{[A]_t} - \frac{1}{[A]_0} = kt$ [10.9a]

$$\frac{1}{56.0\ \text{mmol L}^{-1}} - \frac{1}{220\ \text{mmol L}^{-1}} = k \times 1.22 \times 10^4\ \text{s}$$

$$k = 1.09 \times 10^{-6}\ \text{L mmol}^{-1}\ \text{s}^{-1} = \underline{1.09 \times 10^{-3}\ \text{L mol}^{-1}\ \text{s}^{-1}}$$

11. We have the reaction

$2NO(g) + Cl_2(g) \rightarrow 2NOCl(g)$

This is stated to be a pseudo-second-order reaction in NO, which implies that it is really third-order; second-order in NO, and first-order in Cl_2.

Rate = $kp_{NO}^2 p_{Cl_2} = kp_{Cl_2} \times p_{NO}^2 = k' p_{NO}^2$

where k' is the pseudo-second-order rate constant. Without knowledge of p_{Cl_2}, we cannot calculate k, but we can calculate k'.

at $t = 0$, $p_{NOCl} = 0$, $p_{NO} = 300$ Pa

at $t = 522$s, $p_{NOCl} = 100$ Pa, $p_{NO} = (300 - 100)$ Pa = 200 Pa

$$\frac{1}{p_{NO}} - \frac{1}{p_{NO}(O)} = kt$$

$$\frac{1}{200\ \text{Pa}} - \frac{1}{300\ \text{Pa}} = k \times 522\ \text{s}$$

$k = \underline{3.19 \times 10^{-6}\ \text{Pa}^{-1}\ \text{s}^{-1}}$

13. At $t_{1/2}$ [A] = 1/2[A]$_0$

$(1/2)^n = 1/64$ $\quad$ $n = 6$ half-lives needed

$t = t_{1/2} \times 6 = 221\text{s} \times 6 = \underline{1.33 \times 10^3}$

15. $[^{90}\text{Sr}] = [^{90}\text{Sr}]_0 e^{-kt}$, $k = \dfrac{\ln 2}{t_{1/2}}$

$$k = \frac{\ln 2}{28.1\ \text{y}} = 0.0247\ \text{y}^{-1}$$

Concentrations and masses are proportional to each other, so we may write $m = m_0 e^{-kt}$

(a) $m = 1.00\ \mu\text{g} \times e^{-(0.0247\ \text{y}^{-1} \times 19\ \text{y})} = \underline{0.63\ \mu\text{g}}$

(b) $m = 1.00\ \mu\text{g} \times e^{-(0.0247\ \text{y}^{-1} \times 75\ \text{y})} = \underline{0.16\ \mu\text{g}}$

17. The amount of A in a second order reaction is

$$\frac{1}{[\text{A}]} - \frac{1}{[\text{A}]_0} = kt$$

Therefore

$$t = \frac{1}{k}\left\{\frac{1}{[\text{A}]_t} - \frac{1}{[\text{A}]_0}\right\}$$

$$= \frac{1}{1.24 \times 10^{-3}\ \text{mol}^{-1}\ \text{L s}^{-1}} \times \left\{\frac{1}{0.026\ \text{mol L}^{-1}} - \frac{1}{0.260\ \text{mol L}^{-1}}\right\}$$

$$= \underline{2.8 \times 10^4\ \text{s}}$$

19. $\ln\left(\dfrac{k'}{k}\right) = \dfrac{E_a}{R}\left(\dfrac{1}{T} - \dfrac{1}{T'}\right)$ [10.13]

Solve the above equation for E_a.

$$E_a = \frac{R \ln (k'/k)}{\left(\dfrac{1}{T} - \dfrac{1}{T'}\right)} = \frac{8.315\ \text{J K}^{-1}\ \text{mol}^{-1} \times \ln\left(\dfrac{1.38 \times 10^{-3}}{1.78 \times 10^{-4}}\right)}{\dfrac{1}{292\ \text{K}} - \dfrac{1}{310\ \text{K}}}$$

$$= \underline{85.6\ \text{kJ mol}^{-1}}$$

For A, use

$$A = k \times e^{E_a/RT}$$

$$= 1.78 \times 10^{-4}\ \text{mol L}^{-1}\ \text{s}^{-1} \times e^{85600/8.315 \times 292}$$

$$= \underline{3.66 \times 10^{11}\ \text{mol L}^{-1}\ \text{s}^{-1}}$$

21. In a relative sense, namely the ratio k'/k, the reaction with the greater activation energy is more temperature dependent, that is $\underline{E_a = 52\text{ kJ mol}^{-1}}$.

In an absolute sense $(k' - k)$, the reaction with the smaller activation energy is more temperature dependent, here $\underline{E_a = 25\text{ kJ mol}^{-1}}$. Rate constants are greater for reactions with smaller activation energies.

23. We use the expression for E_a derived in the solution to Exercise 10.19 and assume that the ratio of the rates of spoilage is the ratio of the rate constants.

$$E_a = \frac{R\ln(k'/k)}{\left(\frac{1}{T} - \frac{1}{T'}\right)}$$

$$= \frac{8.315\text{ J K}^{-1}\text{ mol}^{-1} \times \ln(40)}{\frac{1}{277\text{ K}} - \frac{1}{298\text{ K}}} = \underline{120\text{ kJ mol}^{-1}}$$

25. $$E_a = \frac{R\ln(k'/k)}{\left(\frac{1}{T} - \frac{1}{T'}\right)}$$

$$k = \frac{\ln 2}{t_{1/2}} \text{ and } k' = \frac{\ln 2}{2t_{1/2}}$$

$$k'/k = 1/2$$

$$E_a = \frac{8.315\text{ J K}^{-1}\text{ mol}^{-1} \times \ln(1/2)}{\frac{1}{283\text{ K}} - \frac{1}{283\text{ K}}} = \underline{48\text{ kJ mol}^{-1}}$$

27. We use

$$A = P\sigma\left(\frac{8kT}{\pi\mu}\right)^{1/2} N_A \quad \mu = \frac{m_{H_2}m_{C_2H_4}}{m_{H_2} + m_{C_2H_4}} \quad [10.15]$$

$$\sigma(H_2) = 0.27\text{ nm}^2 \quad \sigma(C_2H_4) = 0.64\text{ nm}^2$$

For the effective σ between H_2 and C_2H_4 we will take the average of these two values, that is

$$\sigma = \frac{\sigma(H_2) + \sigma(C_2H_4)}{2} = 0.46\text{ nm}^2$$

$$\mu = \left(\frac{2.016 \times 26.026}{2.016 + 26.026} \right) \text{g mol}^{-1} \Big/ 6.022 \times 10^{23}\ \text{mol}^{-1}$$

$$= 3.11 \times 10^{-24}\ \text{g} = 3.15 \times 10^{-27}\ \text{kg}$$

$p = 1.7 \times 10^{-6}$ (cf p 234)

$$A = 1.7 \times 10^{-6} \times 0.46 \times 10^{-18}\ \text{m}^2 \times \left(\frac{8 \times 1.381 \times 10^{-23}\ \text{J K}^{-1} \times 473\ \text{K}}{\pi \times 3.11 \times 10^{-27}\ \text{kg}} \right)^{1/2} \times 6.022 \times 10^{23}\ \text{mol}^{-1}$$

$$= 1.09 \times 10^{3}\ \text{m}^3\ \text{mol}^{-1}\ \text{s}^{-1}$$

$$= \underline{1.09 \times 10^{6}\ \text{L mol}^{-1}\ \text{s}^{-1}}$$

29. We use

$$k_{\text{rate}} = \left(\frac{kT}{h} e^{\Delta^{\ddagger}S/R} \right) \left(e^{-\Delta^{\ddagger}H/RT} \right) = Ae^{-\Delta^{\ddagger}H/RT}$$

Let us assume that the factor $A = kT/h\ e^{\Delta^{\ddagger}S/R}$ is essentially a constant over this small temperature range (10 K). Then we can determine $\Delta^{\ddagger}H$ from fitting the data to $\ln k_{\text{rate}} = \ln A - \Delta^{\ddagger}H/RT$

k_{rate}	$1.2 \times 10^{-7}\ \text{s}^{-1}$	$4.6 \times 10^{-7}\ \text{s}^{-1}$
$\ln k_{\text{rate}}$	-15.94	-14.59
T	333 K	343 K
$\dfrac{1/T}{\text{K}^{-1}}$	3.003×10^{-3}	2.915×10^{-3}

We find the slope of $\ln k_{\text{rate}}$ against $\left(\frac{1}{T}\right)$ to be

$-1.53 \times 10^{4}\ \text{K} = -\Delta^{\ddagger}H/R$. Then

$\Delta^{\ddagger}H = R \times 1.53 \times 10^{4}\ \text{K} = 1.27 \times 10^{5}\ \text{J mol}^{-1} = 127\ \text{kJ mol}^{-1}$

We also find $\ln A = 29.9$. And

$$\ln A = \ln \left(\frac{kT}{h} \right) + \Delta^{\ddagger}S/R$$

Solving for $\Delta^{\ddagger}S$

$$\Delta^{\ddagger}S = R \ln A + R \ln \left(\frac{h}{kt}\right) \qquad \text{[use average } T = 338 \text{ K]}$$

$$= (8.315 \text{ J K}^{-1} \text{ mol}^{-1} \times 29.9) + \text{R} \ln \left(\frac{6.626 \times 10^{-34}}{1.382 \times 10^{-23} \times 338}\right)$$

$$= 248.6 \text{ J K}^{-1} \text{ mol}^{-1} - 245.96 \text{ J K}^{-1} \text{ mol}^{-1}$$

$$= 2.6 \text{ J K}^{-1} \text{ mol}^{-1}$$

Then $\Delta^{\ddagger}G = \Delta^{\ddagger}H - T\Delta^{\ddagger}S$

$$= 127 \text{ kJ mol}^{-1} - 338 \text{ K} \times 2.6 \text{ J K}^{-1} \text{ mol}^{-1}$$

$$= \underline{126 \text{ kJ mol}^{-1}}$$

31. Rhodopsins from different organisms are likely to be structurally dissimilar, at least in the opsin portion of the molecule, which binds the cis- and trans-retinal molecules. Consequently we might expect that the activation energies for the meta-I to meta-II transformation will be different for bovine rhodopsin and frog rhodopsin. The data suggest that the activation energy in the bovine case is larger than in the frog.

Frogs are cold-blooded creatures, hence it is advantageous to the frog to be able to maintain good vision over a range of temperatures.

Accounting for the Rate Laws

11

1. $K = \frac{k}{k'}$ [11.3]

This relation applies to a second-order reaction as well as to a first-order reaction.
$k = 7.4 \times 10^7\ \text{L mol}^{-1}\ \text{s}^{-1}$

$$k' = \frac{k}{K} = \frac{7.4 \times 10^7\ \text{L mol}^{-1}\ \text{s}^{-1}}{235} = \underline{3.1 \times 10^5\ \text{L mol}^{-1}\ \text{s}^{-1}}$$

3. $A \rightleftharpoons B$

$$\frac{d[A]}{dt} = -k[A] + k'[B] \qquad \frac{d[B]}{dt} = -k'[B] + k[A]$$

$[A] + [B] = [A]_0 + [B]_0$ at all times.

Therefore, $[B] = [A]_0 + [B]_0 - [A]$

$$\frac{d[A]}{dt} = -k[A] + k'\{[A]_0 + [B]_0 - [A]\} = -(k + k')[A] + k'([A]_0 + [B]_0)$$

and $\int_{[A]_0}^{[A]_0} \frac{d[A]}{(k + k')[A] + k'([A]_0 + [B]_0)} = -\int_0^t dt$

The solution is $[A] = \frac{k'([A]_0 + [B]_0) + (k[A]_0 - k'[B]_0)e^{-(k+k')t}}{k + k'}$

Setting $[B]_0 = 0$

$$[A] = \frac{k'[A]_0 + k[A]_0\, e^{-(k+k')t}}{k + k'} = \frac{(k' + ke^{-(k+k')t}[A]_0}{k + k'}$$

which is the expression for [A] in eqn. 11.1. Then

$$[B] = [A]_0 - [A] = \frac{k(1 - e^{-(k+k')t})[A]_0}{k + k'}$$

which agrees with [B] in eqn 11.1.

5. The rate laws are

$$(1)\ \frac{d[A]}{dt} = -k_1[A]$$

$$(2)\ \frac{d[I]}{dt} = -k_1[A] - k_2[I]$$

$$(3)\ \frac{d[P]}{dt} = k_2[I]$$

Differentiating 11.4a yields (1) directly. Differentiating 11.4c yields (3) almost as directly.

$$\frac{d[P]}{dt} = \frac{1}{k_2 - k_1}(-k_1k_2e^{-k_2t} + k_1k_2e^{-k_1t})\ [A]_0$$

$$= \frac{k_1k_2}{k_2 - k_1}(e^{-k_1t} - e^{-k_2t})\ [A]_0$$

$$= k_2[I], \text{ which is (3)}$$

From 11.4b, after differentiating, we have

$$(4)\ \frac{d[I]}{dt} = \frac{k_1}{k_2 - k_1}(-k_1e^{-k_1t} + k_2e^{-k_2t})\ [A]_0$$

The reduction of (4) to (2) requires a little more algebraic manipulation. In this case, it seems easier to work backwards. Thus substitute 11.4a and 11.4b into (2). We obtain

$$\frac{d[I]}{dt} = k_1e^{-k_1t}\ [A]_0 - k_2\left(\frac{k_1}{k_2 - k_1}\right)(e^{-k_1t} - e^{-k_2t})\ [A]_0$$

$$= \left(\frac{k_1}{k_2 - k_1}\right)(k_2 - k_1)e^{-k_1t}\ [A]_0 - \left(\frac{k_1}{k_2 - k_1}\right)k_2(e^{-k_1t} - e^{-k_2t})\ [A]_0$$

$$= \left(\frac{k_1}{k_2 - k_1}\right)\left[(k_2 - k_1)e^{-k_1t}\ [A]_0 - k_2(e^{-k_1t} - e^{-k_2t})\ [A]_0\right]$$

$$= \left(\frac{k_1}{k_2 - k_1}\right)\left[k_2e^{-k_1t}\ [A]_0 - k_1e^{-k_1t}\ [A]_0 - k_2e^{-k_1t}\ [A]_0 + k_2e^{-k_2t}\ [A]_0\right]$$

The first and third terms in the brackets cancel leaving

$$\frac{d[I]}{dt} = \left(\frac{k_1}{k_2 - k_1}\right)(-k_1e^{-k_1t} + k_2e^{-k_2t})\ [A]_0$$

which is (4) above

7. We use equation 11.5 after solving for k_1 and k_2 from the half-lifes

$$k_1 = \frac{\ln 2}{22.5 \text{ d}} = 3.08 \times 10^{-2} \text{ d}^{-1}$$

$$k_2 = \frac{\ln 2}{33.0 \text{ d}} = 2.10 \times 10^{-2} \text{ d}^{-1}$$

$$t = \frac{1}{k_1 - k_2} \ln \frac{k_1}{k_2}$$

$$= \frac{1}{(3.08 - 2.10) \times 10^{-2} \text{ d}^{-1}} \ln (3.08/2.10)$$

$$= \underline{39.1 \text{ d}}$$

9. We assume a pre-equilibrium (that step is fast), and write

$$K = \frac{[\mathrm{A}]^2}{[\mathrm{A}_2]} \text{ so that } [\mathrm{A}] = K^{1/2}[\mathrm{A}_2]^{1/2}$$

The rate-determining step then gives

$$\text{rate} = k_2[\mathrm{A}][\mathrm{B}] = k_2K^{1/2}[\mathrm{A}_2]^{1/2}[\mathrm{B}] = k_{\text{eff}}[\mathrm{A}_2]^{1/2}[\mathrm{B}]$$

where $k_{\text{eff}} = k_2K^{1/2}$

11. We assume that the steady-state approximation applies to [O] (but see the question below). Then

$$\frac{d[\mathrm{O}]}{dt} = 0 = k_1[\mathrm{O}_3] - k_1'[\mathrm{O}][\mathrm{O}_2] - k_2[\mathrm{O}][\mathrm{O}_3]$$

Solving for [O],

$$[\mathrm{O}] = \frac{k_1[\mathrm{O}_3]}{k_1'[\mathrm{O}_2] + k_2[\mathrm{O}_3]}$$

$$\text{Rate} = -\frac{1}{2}\frac{d[\mathrm{O}_3]}{dt}$$

$$\frac{d[\mathrm{O}_3]}{dt} = -k_1[\mathrm{O}_3] + k_1'[\mathrm{O}][\mathrm{O}_2] - k_2[\mathrm{O}][\mathrm{O}_3]$$

Substituting for [O] from above

$$\frac{d[O_3]}{dt} = -k_1[O_3] + \frac{k_1[O_3](k_1'[O_2] - k_2[O_3])}{k_1'[O_2] + k_2[O_3]}$$

$$= \frac{-k_1[O_3](k_1'[O_2] + k_2[O_3]) + k_1[O_3](k_1'[O_2] - k_2[O_3])}{k_1'[O_2] + k_2[O_3]} = \frac{-2k_1k_2[O_3]^2}{k_1'[O_2] + k_2[O_3]}$$

$$\text{Rate} = \frac{k_1k_2[O_3]^2}{k_1'[O_2] + k_2[O_3]}$$

If the second step is slow, then $k_2[O_3] << k_1'[O_2]$ and the rate reduces to

$$\text{Rate} = \frac{k_1k_2[O_3]^2}{k_1'[O_2]}$$

which is second-order in $[O_3]$ and −1 order in $[O_2]$.

Question. Can you determine the rate law expression if the first step of the proposed mechanism is a rapid pre-equilibrium? Under what conditioins does the rate expression above reduce to the case of rapid pre-equilibrium?

13. The rate of production of the product is

$$\frac{d[BH^+]}{dt} = k_2[HAH^+][B]$$

HAH^+ is an intermediate involved in a rapid pre-equilibrium

$$\frac{[HAH^+]}{[HA][H^+]} = \frac{k_1}{k_1'} \text{ so } [HAH^+] = \frac{k_1[HA][H^+]}{k_1'}$$

$$\text{and } \frac{d[BH^+]}{dt} = \frac{k_1k_2}{k_1'}[HA][H^+][B]$$

This rate law can be made independent of $[H^+]$ if the source of H^+ is the acid HA, for then H^+ is given by another equilibrium

$$\frac{[H^+][A^-]}{[HA]} = K_a = \frac{[H^+]^2}{[HA]} \text{ so } [H^+] = (K_a[HA])^{1/2}$$

$$\text{and } \frac{d[BH^+]}{dt} = \frac{k_1k_2K_a^{1/2}}{k_1'}[HA]^{3/2}[B]$$

15. From eqn 11.11 we obtain by substituting $\theta = V/V_{mon}$

$$\frac{V}{V_{mon}} = \frac{p_A}{p_A + K}$$

which can be rewritten as

$$\frac{p_A}{V} = \frac{p_A}{V_{mon}} + \frac{K}{V_{mon}}$$

So a plot of p/V against p has a slope $1/V_{mon}$ and an intercept K/V_{mon}. V_{mon} is calculated from the slope and then K from the intercept.

The data for the plot are as follows:

p/Torr	100	200	300	400	500	600	700
(p/Torr)/(V/cm^3)	9.80	10.8	11.8	12.7	13.6	14.4	15.2

The points are plotted in the figure below.

(b) The slope of the plot is 0.00904 cm^{-3}, so V_{mon} = <u>111 cm^3</u>

(a) The intercept is 8.99 Torr cm^{-3}. Therefore

K = 8.99 Torr cm^{-3} × 111 cm^3 = <u>998 Torr</u>

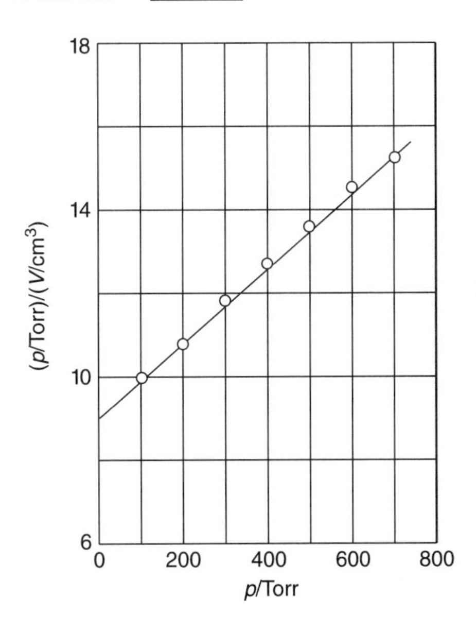

17. (a) We assume that the reaction takes place by encounters between A and B adsorbed on the surface. Therefore for

$A_{ads} + B_{ads} \rightarrow P$ rate $= k\theta_A\theta_B$

$$\theta_A = \frac{K_Ap_A}{1 + K_Ap_A + K_Bp_B} \qquad \theta_B = \frac{K_Bp_B}{1 + K_Ap_A + K_Bp_B}$$

$$\text{rate} = \underline{\frac{kK_AK_Bp_Ap_B}{(1 + K_Ap_A + K_Bp_B)^2}}$$

(b) If p_A and p_B are both very low, we can assume that K_Ap_A and K_Bp_B are both small compared to 1. Then

rate = $\underline{kK_AK_Bp_Ap_B}$

(c) If $p_A = p_B = p$ and if $K_A = K_B = K$, then

$$\text{rate} = \frac{kK^2p^2}{(1 + 2\,Kp)^2}$$

For $2Kp >> 1$, rate $= k$ which is zero order. So yes, this mechanism is consistent with zero-order kinetics at high pressures.

19. Maximum velocity $= K_b[E]_0$ [11.18]

Also, rate $= k[E]_0$ with $k = \dfrac{k_b[S]}{K_M + [S]}$ [11.15]

Therefore,

$$\text{rate} = v = \frac{k_b[S][E]_0}{K_M + [S]} \text{ rearranging}$$

$$k_b[E]_0 = \left\{\frac{K_M + [S]}{[S]}\right\}v$$

$$= \left\{\frac{0.045 \text{ mol L}^{-1} + 0.110 \text{ mol L}^{-1}}{0.110 \text{ mol L}^{-1}}\right\} \times (1.15 \times 10^{-3} \text{ mol L}^{-1} \text{ s}^{-1})$$

$$= \underline{1.62 \times 10^{-3} \text{ mol L}^{-1} \text{ s}^{-1}}$$

21. We start with the Lineweaver-Burk expression, eqn 11.20.

$$\frac{1}{v} = \frac{1}{v_{max}} + \left(\frac{K_M}{v_{max}}\right)\frac{1}{[S]}$$

Multiply both sides of this equation by v/v_{max}.

$$v_{max} = v + K_M\left(\frac{v}{[S]}\right)$$

or

$$v = v_{max} - K_M\left(\frac{v}{[S]}\right)$$

Thus a plot of v against $v/[S]$ yields a straight line with slope equal to $-K_M$ and intercept v_{max}.

23. We fit the data to the Lineweaver-Burk equation [11.20]. Hence draw up the following table after converting to a consistent set of concentration units (mmol L^{-1}).

$(1/[S])/(L\ mmol^{-1})$	1.0	0.50	0.33	0.25	0.20
$(1/v)/(L\ S\ mmol^{-1})$	9.1×10^2	5.6×10^2	4.3×10^2	3.8×10^2	3.4×10^2

A plot of these values is shown in the figure below. This line is close to a straight line with slope 7.0×10^2s and intercept 2.1×10^2 L S mmol^{-1}.
From the intercept

we calculate v_{max}.

$$v_{max} = \frac{1}{\text{intercept}} = \frac{1}{2.1 \times 10^2\ \text{L S mmol}^{-1}} = \underline{4.8 \times 10^{-3}\ \text{mmol L}^{-1}\ \text{s}^{-1}}$$

The slope is K_M/v_{max}, hence

$$K_M = v_{max} \times \text{slope}$$
$$= 4.8 \times 10^{-3}\ \text{mmol L}^{-1}\ \text{s}^{-1} \times 7.0 \times 10^2\text{s}$$
$$= \underline{3.4\ \text{mmol L}^{-1}}$$

The maximum turnover number k_b is given by

$$k_b = \frac{v_{max}}{[E]_0} = \frac{4.8 \times 10^{-3}\ \text{mmol L}^{-1}\ \text{s}^{-1}}{12.5 \times 10^{-3}\ \text{mmol L}^{-1}} = \underline{0.38\ \text{s}^{-1}}$$

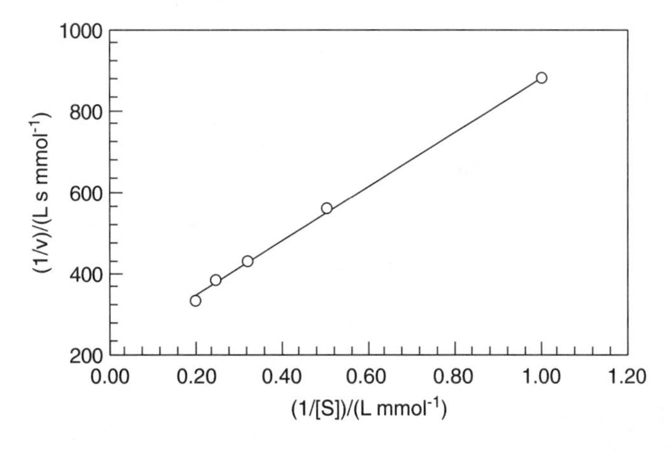

25. We construct Lineweaver-Burk plots for [I] = 0 and for [I] = 15 μmol L^{-1}. If there is a common vertical axis intercept, the inhibition is competitive. If there is a common horizontal axis intercept, the inhibition is noncompetitive.

Draw up the following table.

(a) $\frac{1}{v}/(\mu\text{mol L}^{-1}\text{ s}^{-1})$	2.04	1.05	0.77	0.67	0.62
(b) $\frac{1}{v}/(\mu\text{mol L}^{-1}\text{ s}^{-1})$	3.7	1.92	1.41	1.23	1.16
$\frac{1}{[S]}/(\mu\text{mol L}^{-1})$	100	33	14	8.33	5.56

These data are plotted in the figure below.There is a common intercept on the horizontal axis. Therefore, the inhibition is <u>non-competitive</u>.

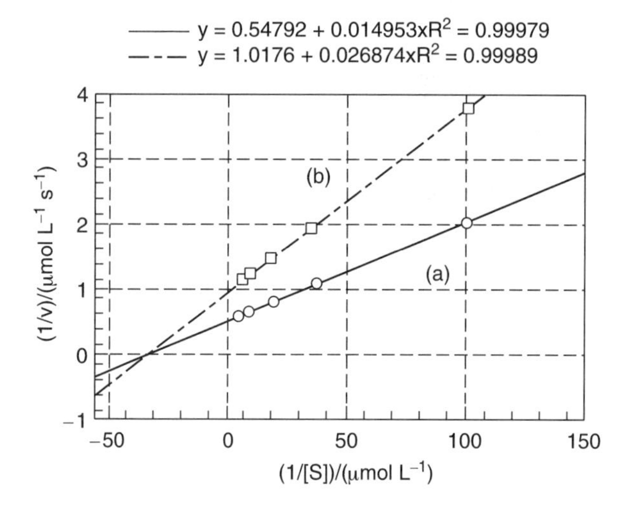

27. $\frac{d[R]}{dt} = 2k_1[R_2] - k_2[R][R_2] + k_3[R'] - 2k_4[R]^2$

$$\frac{d[R']}{dt} = k_2[R][R_2] - k_3[R']$$

Apply the steady-state approximation to both equations

$$2k_1[R_2] - k_2[R][R_2] + k_3[R'] - 2k_4[R]^2 = 0$$

$$k_2[R][R_2] - k_3[R'] = 0$$

The second solves to $[R'] = \frac{k_2}{k_3}[R][R_2]$

and then the first solves to $[R] = \left(\frac{k_1}{k_4}[R_2]\right)^{1/2}$

Therefore, $\frac{d[R_2]}{dt} = -k_1[R_2] - k_2[R_2][R] = \underline{-k_1[R_2] - k_2\left(\frac{k_1}{k_4}\right)^{1/2}[R_2]^{3/2}}$

29. Number of photons absorbed = Φ^{-1} × Number of molecules that react. Therefore,

$$\text{Number absorbed} = \frac{(1.075 \times 10^{-3}\ \text{mol}) \times (6.022 \times 10^{23}\ \text{einstein}^{-1})}{2.1 \times 10^{2}\ \text{mol einstein}^{-1}}$$

$= \underline{3.1 \times 10^{18}}$

Quantum Theory 12

1. We use the Wien displacement law, $T\lambda_{max} = 2.9$ mm K.

$$\lambda_{max} = \frac{2.9 \text{ mm K}}{2773 \text{ K}} = \underline{1.0 \times 10^{-6} \text{ m}}$$

This radiation is in the near infrared region of the spectrum.

3. Assume that the wavelength quoted corresponds roughly to λ_{max}. Then

$$T = \frac{2.9 \text{ mm K}}{\lambda_{max}} = \frac{2.9 \times 10^{-3} \text{ m K}}{6.5 \times 10^{-7} \text{ m}} = \underline{4.5 \times 10^{3} \text{ K}}$$

<u>This radiation is not thermal radiation</u>.

5. $\rho = \dfrac{8\pi hc}{\lambda^5}\left(\dfrac{1}{e^{hc/\lambda kT} - 1}\right)$ [11.5]

As λ increases, $\dfrac{hc}{\lambda kT}$ decreases, and at very long wavelength $hc/\lambda kT \ll 1$.

Hence we can expand the exponential in a power series. Let $x = hc/\lambda kT$, then

$$e^x = 1 + x + \frac{1}{2!}x^2 + \frac{1}{3!}x^3 + \ldots$$

$$\rho = \frac{8\pi hc}{\lambda^5}\left[\frac{1}{1 + x + \frac{1}{2!}x^2 + \frac{1}{3!}x^3 + \ldots - 1}\right]$$

$$\lim_{\lambda \to \infty} \rho = \frac{8\pi hc}{\lambda^5}\left[\frac{1}{1 + x - 1}\right] = \frac{8\pi hc}{\lambda^5}\left(\frac{1}{hc/\lambda kT}\right)$$

$$= \frac{8\pi hc}{\lambda^4}$$

This is the Rayleigh-Jeans law, eqn 11.3

7. We look for the value of λ at which ρ is a maximum, using (as appropriate) the short-wavelength (high-frrequency) approximation

$$\rho = \frac{8\pi hc}{\lambda^5}\left(\frac{1}{e^{hc/\lambda kT}-1}\right) \text{ [12.5]}$$

$$\frac{d\rho}{d\lambda} = -\frac{5}{\lambda}\rho + \frac{hc}{\lambda^2 kT}\left(\frac{e^{hc/\lambda kT}}{e^{hc/\lambda kT}-1}\right)\rho = 0 \text{ at } \lambda = \lambda_{max}$$

Then, $-5 + \dfrac{hc}{\lambda kT} \times \dfrac{e^{hc/\lambda kT}}{e^{hc/\lambda kT}-1} = 0$

Hence, $5 - 5e^{hc/\lambda kT} + \dfrac{hc}{\lambda kT}e^{hc/\lambda kT} = 0$

If $\dfrac{hc}{\lambda kT} >> 1$ [short wavelengths, high frequencies], this expression simplifies.

We neglect the initial 5, cancel the two exponents, and obtain

$hc = 5\lambda kT$ for $\lambda = \lambda_{max}$ and $\dfrac{hc}{\lambda kT} >> 1$

or $\lambda_{max}T = \dfrac{hc}{5k}$ = 2.88 mm K, in accord with observation.

9. $\Delta E = h\omega = h\nu = \dfrac{h}{t}$ $\left[T = \text{period} = \dfrac{1}{\nu} = \dfrac{2\pi}{\omega}\right]$

(a) $\Delta E = 6.626 \times 10^{-34}$ J s $\times 1.0 \times 10^{15}$ s^{-1} $= 6.6 \times 10^{-19}$ J corresponding to $N_A \times 6.6 \times 10^{-19}$ J = 4.0×10^2 kJ mol^{-1}

(b) $\Delta E = \dfrac{6.626 \times 10^{-34}\text{ J s}}{2.0 \times 10^{-14}\text{ s}} = 3.3 \times 10^{-20}$ J = 20 kJ mol^{-1}

(c) $\Delta E = \dfrac{6.626 \times 10^{-34}\text{ J s}}{0.50\text{ s}} = 1.3 \times 10^{-33}$ J = 8.0×10^{-13} kJ mol^{-1}

11. $P = \dfrac{E}{t}$

$N = \dfrac{P}{h\nu} = \dfrac{P\lambda}{hc}$ [P = power in J s^{-1}]

$$= \frac{P\lambda}{(6.626 \times 10^{-34}\text{ J Hz}^{-1}) \times (2.998 \times 10^8\text{ m s}^{-1})}$$

$$= \frac{(\mathrm{P/W}) \times (\lambda/\mathrm{nm})\mathrm{s}^{-1}}{1.99 \times 10^{-16}} = 5.03 \times 10^{15}(\mathrm{P/W}) \times (\lambda/\mathrm{nm})\ \mathrm{s}^{-1}$$

(a) $N = (5.03 \times 10^{15}) \times 1.0 \times 350\ \mathrm{s}^{-1} = \underline{1.8 \times 10^{18}\ \mathrm{s}^{-1}}$

(b) $N = (5.03 \times 10^{15}) \times 100 \times 550\ \mathrm{s}^{-1} = \underline{1.8 \times 10^{20}\ \mathrm{s}^{-1}}$

13. $\Phi = 2.14\ \mathrm{eV} = 2.14\ \mathrm{eV} \times 1.602 \times 10^{-19}\ \mathrm{J\ eV}^{-1}$

$= 3.43 \times 10^{-19}\ \mathrm{J}$

$$\frac{1}{2} m_e v^2 = h\nu - \Phi = \frac{hc}{\lambda} - \Phi\ [12.7]$$

(a) $$\frac{hc}{\lambda} = \frac{(6.626 \times 10^{-34}\ \mathrm{J\ s}) \times 2.998 \times 10^{8}\ \mathrm{m\ s}^{-1}}{(750 \times 10^{-9}\ \mathrm{m})}$$

$= 2.65 \times 10^{-19}\ \mathrm{J} < \Phi$, Therefore, no ejection occurs.

(b) $$\frac{hc}{\lambda} = \frac{(6.626 \times 10^{-34}\ \mathrm{J\ s}) \times 2.998 \times 10^{8}\ \mathrm{m\ s}^{-1}}{(250 \times 10^{-9}\ \mathrm{m})}$$

$= 7.95 \times 10^{-19}\ \mathrm{J}$

Hence $1/2\ mv^2 = (7.95 - 3.43) \times 10^{-19}\ \mathrm{J} = \underline{4.52 \times 10^{-19}\ \mathrm{J}}$

$$v = \left(\frac{2 \times 4.52 \times 10^{-19}\ \mathrm{J}}{9.109 \times 10^{-31}\ \mathrm{kg}}\right)^{1/2} = \underline{996\ \mathrm{km\ s}^{-1}}$$

15. $$\lambda = \frac{h}{p} = \frac{h}{mv}$$

(a) $$\lambda = \frac{(6.626 \times 10^{-34}\ \mathrm{J\ s})}{(1.00\ \mathrm{m\ s}^{-1}) \times 1.0 \times 10^{-3}\ \mathrm{kg}} = \underline{6.6 \times 10^{-31}\ \mathrm{m}}$$

(b) $$\lambda = \frac{(6.626 \times 10^{-34}\ \mathrm{J\ s})}{(1.0 \times 10^{8}\ \mathrm{m\ s}^{-1}) \times 1.0 \times 10^{-3}\ \mathrm{kg}} = \underline{6.6 \times 10^{-39}\ \mathrm{m}}$$

(c) $$\lambda = \frac{(6.626 \times 10^{-34}\ \mathrm{J\ s})}{4.003 \times (1.6605 \times 10^{-27}\ \mathrm{kg}) \times (1.0 \times 10^{3}\ \mathrm{m\ s}^{-1})} = \underline{99.7\ \mathrm{pm}}$$

17. $p = \dfrac{h}{\lambda}$

(a) $p = \dfrac{(6.626 \times 10^{-34}\ \text{J s})}{(725 \times 10^{-9}\ \text{m})} = \underline{9.14 \times 10^{-28}\ \text{kg m s}^{-1}}$

(b) $p = \dfrac{(6.626 \times 10^{-34}\ \text{J s})}{(75 \times 10^{-12}\ \text{m})} = \underline{8.8 \times 10^{-24}\ \text{kg m s}^{-1}}$

(c) $p = \dfrac{(6.626 \times 10^{-34}\ \text{J s})}{(20\ \text{m})} = \underline{3.3 \times 10^{-35}\ \text{kg m s}^{-1}}$

19. $p = \dfrac{h}{\lambda} = \dfrac{6.626 \times 10^{-34}\ \text{J s}}{\lambda} = mv$

$$v = \frac{(6.626 \times 10^{-34}\ \text{J s})}{(300 \times 10^{-9}\ \text{m}) \times 1.0 \times 10^{-3}\ \text{kg}}$$

$= \underline{2.2 \times 10^{-24}\ \text{m s}^{-1}}$

21. This is essentially the photoelectric effect with the work function Φ being the ionization energy I. Hence,

$$\frac{1}{2} m_e v^2 = h\nu - I = \frac{hc}{\lambda} - I$$

Solving for λ

$$\lambda = \frac{hc}{I + \frac{1}{2} mv_2} = \frac{(6.626 \times 10^{-34}\ \text{J s}) \times (2.998 \times 10^{8}\ \text{m s}^{-1})}{(3.44 \times 10^{-18}\ \text{J}) + \left(\frac{1}{2}\right) \times (9.109 \times 10^{-31}\ \text{kg}) \times (1.03 \times 10^{6}\ \text{m s}^{-1})^2}$$

$= 5.06 \times 10^{-8}\ \text{m} = \underline{50.6\ \text{nm}}$

Question. What is the energy of the photon?

23. The required probability is

$$P = \int_{L_1}^{L_2} \psi^2 \text{d}x = \int_{L_1}^{L_2} \left(\frac{2}{L}\right) \sin^2\left(\frac{\pi x}{L}\right) \text{d}x$$

The necessary integral can be found in tables of integrals and is

$$\frac{1}{2} x - \frac{1}{4a} \sin 2ax \left[a = \frac{\pi}{L}\right]$$

The definite integrals involved can also be evaluated with mathematical software, such as MathCad, or with hand-held calculators. Approximate answers can be obtained by assuming that ψ is constant over the relatively small range of x.

(a) $L_2 = \frac{0.2}{10} L \qquad L_1 = \frac{0.1}{10} L$

The integral evaluates to $P = \underline{4.6 \times 10^{-5}}$

(b) $L_2 = \frac{5.2}{10} L \qquad L_1 = \frac{4.9}{10} L$

$P = \underline{6.0 \times 10^{-2}}$

These values were obtained with MathCad 8.

25. $\Delta p \Delta x \geq \frac{1}{2} \hbar \qquad \Delta p = m \Delta v$

$$\Delta v_{\text{min}} = \frac{\hbar}{2m\Delta x} = \frac{1.055 \times 10^{-34}\ \text{J s}}{2 \times 0.500\ \text{kg} \times 5.0 \times 10^{-6}\ \text{m}} = \underline{2.1 \times 10^{-29}\ \text{m s}^{-1}}$$

27. The minimum uncertainty in position is 100 pm. Therefore, because $\Delta x \Delta p \geq \frac{1}{2} \hbar$

$$\Delta p \geq \frac{\hbar}{2\Delta x} = \frac{1.0546 \times 10^{-34}\ \text{J s}}{2(100 \times 10^{-12}\ \text{m})} = 5.3 \times 10^{-25}\ \text{kg m s}^{-1}$$

$$\Delta v = \frac{\Delta p}{m} = \frac{5.3 \times 10^{-25}\ \text{kg m s}^{-1}}{9.11 \times 10^{-31}\ \text{kg}} = \underline{5.8 \times 10^{-5}\ \text{m s}^{-1}}$$

29. The maximum probability occurs at $\frac{\pi x}{L} = \frac{\pi}{2}$, corresponding to $x = 1/2\ L$.

At that location

$$P_{\text{max}}\ \alpha\ \psi^2 = \left(\frac{2}{L}\right) \sin^2\left(\frac{\pi}{2}\right) = \frac{2}{L}$$

$$50\%\ \text{max} = 0.5\ \frac{2}{L} = \frac{1}{L}$$

We need to find the location at which $P_{1/2} = \frac{1}{L}$. This corresponds to the angle at which $\sin^2\theta = \frac{1}{2}$. This is $\theta = \frac{\pi}{4}$, corresponding to $x = \frac{L}{4}$ or $x = \frac{3L}{4}$

31. $$\int_0^\infty \psi^2 dx = \int_0^L \psi^2 dx = \int_0^L A^2 dx = 1$$

$$= A^2 x \Big|_0^L = A^2 L = 1$$

Therefore $A^2 = \frac{1}{L}$, $A = \left(\frac{1}{L}\right)^{1/2}$

The normalized wave function is $\underline{\psi = \left(\frac{1}{L}\right)^{1/2}}$

33. (a) $I = mr^2$

$= 1.008 \text{ u} \times 1.6605 \times 10^{-27} \text{ kg/u} \times (1.61 \times 10^{-10} \text{ m})^2$

$= \underline{4.34 \times 10^{-47} \text{ kg m}^2}$

(b) $E = \dfrac{m_l^2 \hbar^2}{2I}$ [12.16]

$E_0 = 0$ [$m_l = 0$]

$E_1 = \dfrac{\hbar^2}{2I}$ [$m_l = 1$]

$$\Delta E = E_1 - E_0 = h\nu = \frac{hc}{\lambda} = \frac{\hbar^2}{2I} = \frac{h^2}{8\pi^2 I}$$

$$\lambda = \frac{8\pi^2 cI}{h} = \frac{8\pi^2 \times 2.998 \times 10^8 \text{ m s}^{-1} \times 4.34 \times 10^{-47} \text{ kg m}^2}{6.626 \times 10^{-34} \text{ J s}}$$

$\lambda = 1.55 \times 10^{-3} \text{ m} = \underline{1.55 \text{ mm}}$

This wavelength is in the microwave region of the electromagnetic spectrum.

35. (a) $\nu = \dfrac{1}{2\pi}\left(\dfrac{k}{m}\right)^{1/2}$ [12.19]

$$= \frac{1}{2\pi}\left(\frac{314 \text{ N m}^{-1}}{1.008 \text{ u} \times 1.6605 \times 10^{-27} \text{ kg/u}}\right)^{1/2} = \underline{6.89 \times 10^{13} \text{ s}^{-1}}$$

(b) $\lambda = \dfrac{c}{\nu} = \dfrac{2.998 \times 10^8 \text{ m s}^{-1}}{2.42 \times 10^{13} \text{ s}^{-1}} = 4.35 \times 10^{-6} \text{ m} = \underline{4.35\ \mu\text{m}}$

Atomic Structure

1. $\tilde{\nu} = \dfrac{1}{\lambda} = \mathcal{R}_{\mathrm{H}}\left(\dfrac{1}{n_1^2} - \dfrac{1}{n_2^2}\right)$ [13.1]

and $n_1 = 2, n_2 = 5$

$\mathcal{R}_{\mathrm{H}} = 109\,677\ \mathrm{cm^{-1}} = 1.09677 \times 10^7\ \mathrm{m^{-1}}$

Solving for λ

$\lambda = (1.09677 \times 10^7\ \mathrm{m^{-1}})^{-1}\left(\dfrac{1}{4} - \dfrac{1}{25}\right)^{-1}$

$= \underline{434\ \mathrm{nm}}$

3. $\dfrac{1}{\lambda} = \dfrac{1}{486.1 \times 10^{-7}\ \mathrm{cm}} = 20572\ \mathrm{cm^{-1}}$

Hence, the term lies at

$27414\ \mathrm{cm^{-1}} - 20572\ \mathrm{cm^{-1}} = \underline{6842\ \mathrm{cm^{-1}}}$

(b) $E = \dfrac{hc}{\lambda} = (6.626 \times 10^{-34}\ \mathrm{J\ s}) \times 2.998 \times 10^8\ \mathrm{m\ s^{-1}} \times 6842\ \mathrm{cm^{-1}} \times 100\ \mathrm{m^{-1}/cm^{-1}}$

$E = \underline{1.36 \times 10^{-19}\ \mathrm{J}}$

5. Examine eqn 13.4 and see that

$\Delta E \propto Z^2\left(\dfrac{1}{n_1^2} - \dfrac{1}{n_2^2}\right) \propto \nu$

We look for $n_1(\mathrm{He^+})$ and $n_2(\mathrm{He^+})$ that satisfies

$\left(\dfrac{1}{1^2} - \dfrac{1}{2^2}\right) = 4\left(\dfrac{1}{n_1^2(\mathrm{He^+})} - \dfrac{1}{n_2^2(\mathrm{He^+})}\right)$

clearly $n_1(\mathrm{He^+}) = 2n_1(\mathrm{H}) = 2$

$n_2(\mathrm{He^+}) = 2n_2(\mathrm{H}) = 4$

An allowed transition would be $\underline{4p \rightarrow 2s}$ in $\mathrm{He^+}$.

7. $n = 4$ for the N shell

$n^2 = 4^2 = 16$ orbitals

9. λ_{max} (H) = 12368 nm

$$\Delta E = h\nu = \frac{hc}{\lambda}$$

$\Delta E(\mathrm{He}^+) \approx Z^2 \Delta E(\mathrm{H}) = 4\Delta E_{\mathrm{H}}$; therefore

$$\lambda_{max}(\mathrm{He}^+) \approx \frac{1}{4}\lambda_{max}(\mathrm{H})$$

$$= \frac{12368 \text{ nm}}{4} = \underline{3092 \text{ nm}}$$

11. A Lyman series corresponds to $n_1 = 1$; hence

$$\tilde{\nu} = \mathcal{R}_{\mathrm{Li}^{2+}}\left(1 - \frac{1}{n_2}\right), n = 2, 3, \ldots \left[\tilde{\nu} = \frac{1}{\lambda}\right]$$

Therefore, if the formula is appropriate, we expect to find that

$\tilde{\nu}\left(1 - \frac{1}{n_2}\right)^{-1}$ is a constant ($\mathcal{R}_{\mathrm{Li}^{2+}}$). We therefore draw up the following table.

n	2	3	4
$\tilde{\nu}/\mathrm{cm}^{-1}$	740 747	877 924	925 933
$\tilde{\nu}\left(1 - \frac{1}{n_2}\right)^{-1}/\mathrm{cm}^{-1}$	987 663	987 665	987 662

Hence, the formula does describe the transitions, and $\underline{\mathcal{R}_{\mathrm{Li}^{2+}} = 987\ 663 \text{ cm}^{-1}.}$

The Balmer transitions lie at

$$\tilde{\nu} = \mathcal{R}_{\mathrm{Li}^{2+}}\left(\frac{1}{4} - \frac{1}{n^2}\right) \qquad n = 3, 4, \ldots$$

$$= (987\ 663 \text{ cm}^{-1}) \times \left(\frac{1}{4} - \frac{1}{n^2}\right) = \underline{137\ 175 \text{ cm}^{-1}, 185\ 187 \text{ cm}^{-1}}, \ldots$$

The ionization energy of the ground-state ion is given by

$$\tilde{\nu} = \mathcal{R}_{\mathrm{Li}^{2+}}\left(1 - \frac{1}{n_2}\right), \qquad n \to \infty$$

and hence corresponds to

$\tilde{\nu} = 987\ 663\ \mathrm{cm}^{-1}$, or 122.5 eV

13. The radial distribution function varies as

$$P = 4\pi r^2 \psi^2 = \frac{4r^2}{a_0^3}\,\mathrm{e}^{-2r/a_0}$$

The maximum value of P occurs at $r = a_0$ because

$$\frac{\mathrm{d}P}{\mathrm{d}r} \propto \left(2r - \frac{2r^2}{a_0}\right)\mathrm{e}^{-2r/a_0} = 0 \text{ at } r = a_0 \text{ and } P_{\mathrm{max}} = \frac{4}{a_0}\,\mathrm{e}^{-2}$$

P falls to a fraction f of its maximum when

$$f = \frac{\dfrac{4r^2}{a_0^3}\,\mathrm{e}^{-2r/a_0}}{\dfrac{4}{a_0}\,\mathrm{e}^{-2}} = \frac{r^2}{a_0^2}\,\mathrm{e}^2\mathrm{e}^{-2r/a_0}$$

Therefore solve

$$\frac{f^{1/2}}{\mathrm{e}} = \left(\frac{r}{a_0}\right)\mathrm{e}^{-r/a_0}$$

(a) $f = 0.25$

solves to $r = 0.7569\ a_0$ or $0.2431\ a_0$ = 40 pm or 13 pm

(b) $f = 0.10$

solves to $r = 0.554\ a_0$ or $0.446\ a_0$ = 29 or 24 pm

15. $V = 5.0\ \mathrm{pm}^3 = \frac{4}{3}\pi r^3$ [assume a spherical volume]

$$r = \sqrt[3]{\frac{3V}{4\pi}} = \sqrt[3]{\frac{3 \times 5.0 \times 10^{-36}\ \mathrm{m}^3}{4\pi}} = 1.06 \times 10^{-12}\ \mathrm{m}$$

$$= 1.06\ \mathrm{pm}$$

As the most probable radius of the electron in the ground state of the H-atom is $a_0 = 52.9$ pm, it is probably safe to assume that $\psi^2(r)$ is a constant within the spherical volume with $r = 1.06$ pm, Therefore,

$$\psi^2(r) = \frac{1}{\pi a_0^3}\, e^{-2r/a_0} = \text{constant}$$

Evaluate $\psi^2(r)$ at $r = 0.53$ pm, which is an average value of r within the volume.

$$\psi^2 = \frac{1}{\pi(53\text{ pm})^3} e^{-0.02} = \frac{0.98}{\pi(53\text{ pm})^3}$$

Then,

$$\text{Probability} = \int \psi^2(r)\delta V \approx \psi^2\,\delta V$$

$$= \frac{0.98}{\pi(53\text{ pm})^3} \times 5.0\text{ pm}^3 = \underline{1.1 \times 10^{-5}}$$

17. $P(r) = 4\pi r^2\psi^2 = \dfrac{r^2}{8a_0^3}\left(2 - \dfrac{2r}{a_0}\right)^2 e^{-r/a_0}$

To create the plot express r in units of a_0, then $P(r)/8a_0 = r^2(2 - r)^2 e^{-r}$

This function is plotted in the figure below.

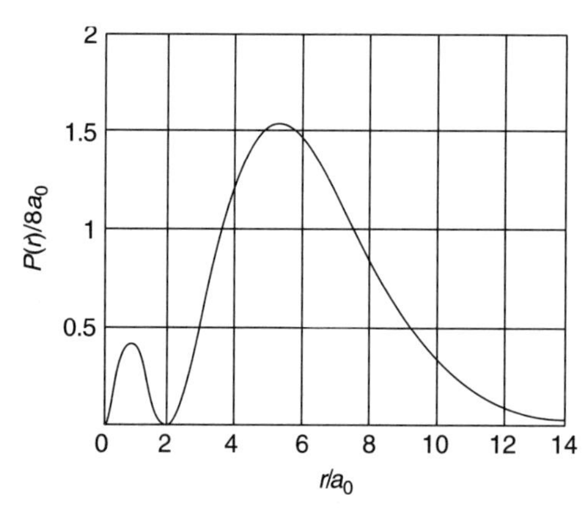

An examination of the graph indicates that the most probable value of $r \approx \underline{5.2\ a_0}$. Differentiating $P(r)/8a_0$ yields the following maxima and minima in terms of r/a_0.

minima: 0, 2

maxima: $3 - \sqrt{5}$, $\underline{3 + \sqrt{5}}$

These values are clearly consistent with the graph in the figure above.

19. Look for values of θ for which sin θ or cos θ go to zero. Sin θ goes to zero at $\theta = 0°$ and $180°$; cos θ at $90°$ and $270°$.

21. The energies are $E = -\dfrac{hc\mathcal{R}_H}{n^2}$, and the orbital degeneracy g of an energy level of principal quantum number n is $g = n^2$

   (a) $E = -hc\mathcal{R}_H$ implies that $n = 1$, so $g = 1$ (the 1s orbital)

   (b) $E = -\dfrac{hc\mathcal{R}_H}{9}$ implies that $n = 3$, so $g = 9$ (the 3s orbital, the three p orbitals, and the five 3d orbitals)

   (c) $E = -\dfrac{hc\mathcal{R}_H}{49}$ implies that $n = 7$, so $g = 49$ (the 7s orbital, the three 7p orbitals, the five 7d orbitals, the seven 7f orbitals, the nine 7g orbitals).

23. All d and g orbitals

25. Use the building-up principle with the orbitals occupied in the order 1s, 2s, 2p, 3s, 3p

H					He
$1s^1$					$1s^2$
Li	Be	B	C	N	O
$[He]2s^1$	$[He]2s^2$	$[He]2s^22p^1$	$[He]2s^22p^2$	$[He]2s^2sp^3$	$[He]2s^2sp^4$
				F	Ne
				$[He]2s^22p^5$	$[He]2s^22p^6$
Na	Mg	Al	Si	P	S
$[Ne]3s^1$	$[Ne]3s^2$	$[Ne]3s^23p^1$	$[Ne]3s^23p^2$	$[Ne]3s^23p^3$	$[Ne]3s^23p^4$
				Cl	Ar
				$[Ne]3s^23p^5$	$[Ne]3s^23p^6$

Where $[He] = 1s^2$, $[Ne] = 2s^22p^6$

27. (a) Periodic Table

1s	H	He										
2s	Li	Be	B	C	N	O	F	Ne	Na	Mg	2p
3s	Al	Si	P	S	Cl	Ar	K	Ca	Sc	Ti	3p
4s	V	Cr									

↑ "Noble Gases" (Mg, Ti)

29. The coupling of a p electron ($l = 1$) and a d electron ($l = 2$) gives rise to $L = 3$ (F), 2 (D), and 1 (P) terms. Possible values of S include 0 and 1. Possible values of J (using Russell-Saunders coupling) are 3, 2, and 1 ($S = 0$) and 4, 3, 2, 1, and 0 ($S = 1$). The term symbols are

$\underline{{}^1F_3;\ {}^3F_4,\ {}^3F_3,\ {}^3F_2;\ {}^1D_2;\ {}^3D_3,\ {}^3D_2,\ {}^3D_1;\ {}^1P_1;\ {}^3P_2,\ {}^3P_1,\ {}^3P_0.}$

Hund's rules state that the lowest energy level has maximum multiplicity. Consideration of spin–orbit coupling says the lowest energy level has the lowest value of $J(J + 1) - L(L + 1) - S(S + 1)$. So the lowest energy level is $\underline{{}^3F_2}$.

31. (a) $L = 0, S = 0, J = 0$ $\underline{1}$ level

(b) $L = 3, S = 1, J = 4, 3, 2$ $\underline{3}$ levels

(c) $L = 0, S = 2, J = 2$ $\underline{1}$ level

(d) $L = 1, S = 2, J = 3, 2, 1$ $\underline{3}$ levels

The Chemical Bond 14

1. The valence bond description of P_2 is similar to that of N_2: One $\sigma(2p_{zA}, 2p_{zB})$; one $\pi(2p_{xA}, 2p_{xB})$; and one $\pi(2p_{yA}, 2p_{yB})$ bond; along with their antibonding counterparts.

   In the tetrahedral P_4 molecule there are six single P—P bonds of roughly 200 kJ mol^{-1} bond enthalpy each. So the total bonding enthalpy is roughly 1200 kJ mol^{-1}. In the transformation

   $$P_4 \rightarrow 2P_2$$

   there is a loss of about 800 kJ mol^{-1} in σ-bond enthalpy. This loss is not likely to be made up by the formation of 4 P—P π-bonds. Period 3 atoms, such as P, are too large to get close enough to each other to form strong π-bonds.

3. We use eqn 14.3 with R = 74.1 pm

   $$V_{\text{nuc, nuc}} = \frac{e^2}{4\pi\varepsilon_0 R} = \frac{(1.602 \times 10^{-19}\text{ C})^2}{1.113 \times 10^{-10}\text{ J}^{-1}\text{ C}^2\text{ m}^{-1} \times 74.1 \times 10^{-12}\text{ m}}$$

   $$= 3.11 \times 10^{-18}\text{ J}$$

   The molar value is

   $6.022 \times 10^{23}\text{ mol}^{-1} \times 3.11 \times 10^{-18}\text{ J} = \underline{1.87 \times 10^{6}\text{ J mol}^{-1}}$

5. See Further Information 9 and 10 for a summary of the methods of determining Lewis structures and how they are used in the VSEPR model to obtain the geometries of molecules.

   (a) CO_2 is linear, either by VSEPR theory (two atoms attached to the central atom, no lone pairs on C) or by regarding the molecule as having a σ framework and π-bonds between the C and O atoms.

   (b) NO_2 is angular because it is isoelectronic with CO_2^-. The extra electron is a "half lone pair" and a bending agent. The extra electron can be accommodated by the molecule bending so as to give the lone pair some s orbital character.

   (c) NO_2^+ is linear because it is isoelectronic with CO_2.

   (d) NO_2^- is angular because it has one more electron than NO_2 and a correspondingly stronger bending influence.

(e) SO_2 is angular, its valence electron structure is similar to NO_2^-.

(f) H_2O is angular, there are 4 electron pairs around the O atom resulting in the O—H bonds being roughly at the tetrahedral angle.

(g) H_2O_2 is angular around each O atom for the same reason as in H_2O.

7.

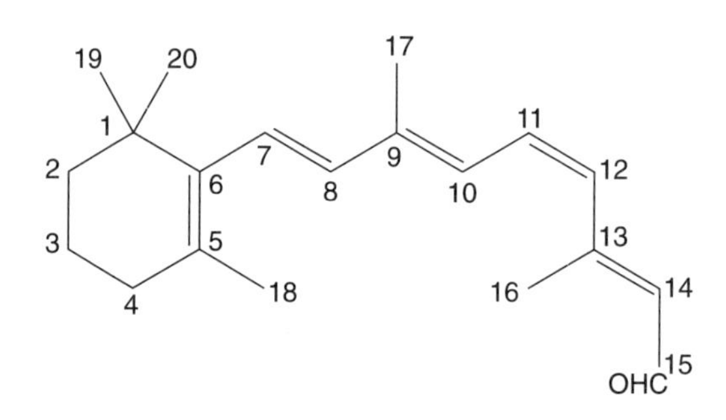

Refer to the structure above for the numbering of the carbon atoms in cis-retinal. Carbon atoms 5-15 each have three sp^2 hybrid atomic orbitals which form σ-bonds with their neighboring atoms. There are six conjugated π-bonds between these 11 C atoms and the one O atom. These six π-bonds are formed from 12 p_x atomic orbitals, one on each of the 12 atoms. They are resonance hybrids all of the form:

$$\psi(\pi\text{-bond}) = \sum_{i=5}^{15} \psi_{2p_xC_i} + \psi_{2p_xO}$$

All the remaining C atoms each have four sp^3 hybrid atomic orbitals which form σ-bonds with their neighboring atoms.

9. We need to demonstrate that $\int \psi^2 d\tau = 1$, where $\psi = \dfrac{s + \sqrt{2}\,p}{\sqrt{3}}$.

$$\int \psi^2 d\tau = \frac{1}{3}\int (s + \sqrt{2}\,p)^2\, d\tau = \frac{1}{3}\int (s^2 + 2p^2 + 2\sqrt{2}\,sp)^2\, d\tau = \frac{1}{3}(1 + 2 + 0) = 1$$

as $\int s^2 d\tau = 1$, $\int p^2 d\tau = 1$, and $\int sp\, d\tau = 0$ [orthogonality]

11. $\int \psi^2 d\tau = N^2 \int (\psi_{\text{cov}} + \lambda_{\text{ion}})^2 d\tau = 1$

$$= N^2 \int (\psi_{\text{cov}}^2 + \lambda^2 \psi_{\text{ion}}^2 + 2\lambda\psi_{\text{cov}}\psi_{\text{ion}}) d\tau = 1$$

$$= N^2(1 + \lambda^2 + 2\lambda S)$$

Where we have assumed that ψ_{cov} and ψ_{ion} are individually normalized and we have written

$$S = \int \psi_{\text{cov}} \psi_{\text{ion}} \, \mathrm{d}\tau$$

$$\text{Hence } N = \left(\frac{1}{1 + 2\lambda S + \lambda^2} \right)^{1/2}$$

13. Covalent structures:

Ionic structures:

In addition there are many other possible ionic structures. These structures can be safely ignored in simple descriptions of the molecule because the coefficients of the wave function representing these structures in the linear combination of wave functions for the entire resonance hybrid are very small. Benzene is a very symmetrical molecule, and we expect that all the C atoms will be equivalent. Hence, those structures in which the C atoms are not equivalent should contribute little to the resonance hybrid.

15. σ-bonding with $d_{x^2-y^2}$ orbital

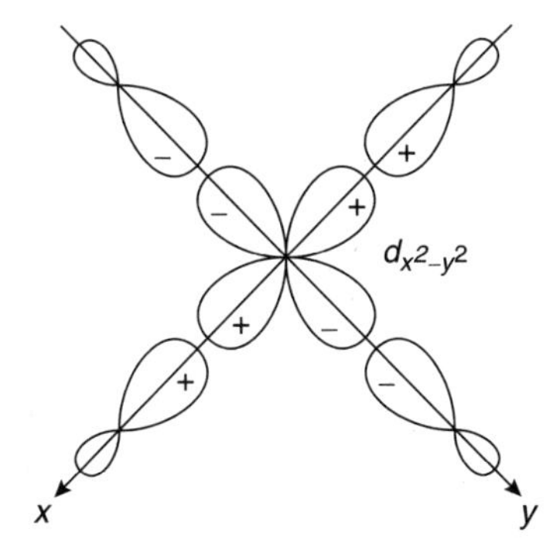

The σ-antibonding orbital looks the same but with the p-orbital lobes pointed in the opposite direction.

The σ-bonding and antibonding diagrams with the d_{xy}, d_{yz}, and d_{xz} orbitals have the same appearance as the diagram above except that the *d*-orbital lobes are pointed between the axes rather than along them.

σ-bonding with d_{yz} orbital

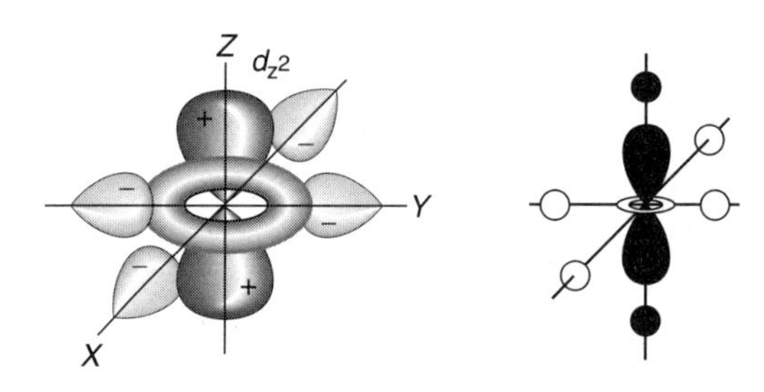

In this figure, only one of the *p*-orbital lobes is shown in each *p*-orbital. *p*-orbitals with positive lobes may also approach this orbital along the + and – *z*-direction as indicated in the smaller diagram on the right. The antibonding diagrams are similar, but with the signs of the *p*-orbital lobes reversed.

π-bonding

Only the d_{xy}, d_{yz}, and d_{xz} orbitals undergo π-bonding with *p*-orbitals on neighboring atoms. The bonding arrangement is pictured below with the d_{xy} orbital. The diagrams for the d_{yz} and d_{xz} orbitals are similar. The antibonding diagrams have the signs of the *p*-orbital lobes reversed.

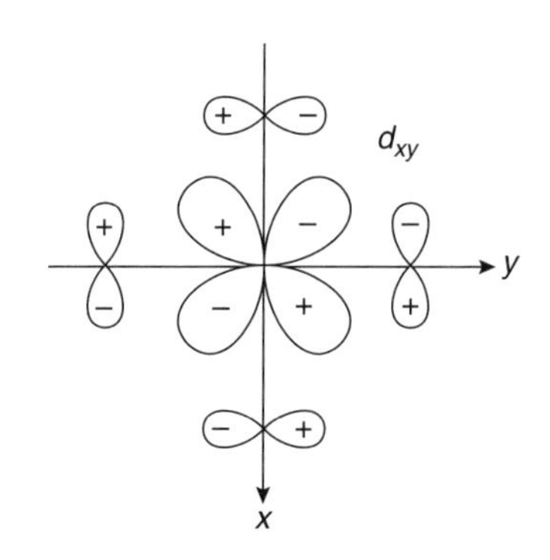

17. (a) H_2^- $1\sigma^2 2\sigma^{*1}$ $b = \frac{1}{2}$

(b) N_2 $1\sigma^2 2\sigma^{*2} 1\pi^4 3\sigma^2$ $b = 3$

(c) O_2 $1\sigma^2 2\sigma^{*2} 3\sigma^2 1\pi^4 2\pi^{*2}$ $b = 2$

19. B_2 $\quad 1\sigma^2 2\sigma^{*2} 1\pi^2 \quad b = 1$

C_2 $\quad 1\sigma^2 2\sigma^{*2} 1\pi^4 \quad b = 2$

The bond orders of B_2 and C_2 are respectively 1 and 2. Therefore $\underline{C_2}$ should have the greater bond dissociation enthalpy. The experimental values are approximately 4 eV and 6 eV respectively.

21. We use Figure 14.18 of the text.

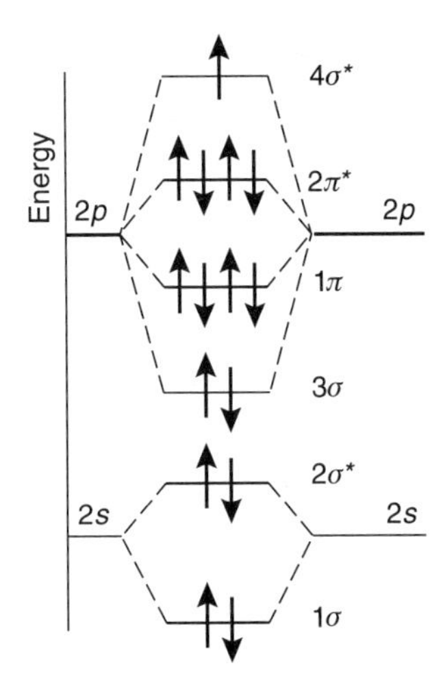

Because the bond order is increased when XeF^+ is formed from XeF (an electron is removed from an antibonding orbital), $\underline{XeF^+ \text{ will have a shorter bond length than XeF}}$.

23. The wavefunctions are

$$\psi_n = \left(\frac{2}{L}\right)^{1/2} \sin\left(\frac{n\pi x}{L}\right)$$

The center is at $\frac{L}{2}$. We invert through $x = \frac{L}{2}$. For $n = 1$, when $x < \frac{L}{2}$,

$\psi_1 = +$, when $x > \frac{L}{2}$, $\psi_1 = +$, therefore ψ_1 is $\underline{g}$. In a similar fashion we determine

$n = 2$, $\underline{u}$

$n = 3$, $\underline{g}$

$n = 4$, $\underline{u}$

25. The parities are given in the figure. a_{2u}, e_{1g}, e_{2u}, and b_{2g}.

27. F_2^+ $\quad 1\sigma^2 2\sigma^{*2} 3\sigma^2 1\pi^4 2\pi^{*3}$ $\quad b = \frac{3}{2}$

F_2 $\quad 1\sigma^2 2\sigma^{*2} 3\sigma^2 1\pi^4 2\pi^{*4}$ $\quad b = 1$

F_2^- $\quad 1\sigma^2 2\sigma^{*2} 3\sigma^2 1\pi^4 2\pi^{*4} 4\sigma^{*1}$ $\quad b = \frac{1}{2}$

Therefore, order of bond lengths is $F_2^+ < F_2 < F_2^-$.

29. We use the molecular orbital energy level diagram in Fig. 14.32. As usual, we fill the orbitals starting with the lowest energy orbital, obeying the Pauli principle and Hund's rule. We then write

(a) $C_6H_6^-$ (7 electrons): $\underline{a_{2u}^2 e_{1g}^4 e_{2u}^1}$

$E = 2(\alpha + 2\beta) + 4(\alpha + \beta) + (\alpha - \beta) = \underline{7\alpha + 7\beta}$

(b) $C_6H_6^+$ (5 electrons): $\underline{a_{2u}^2 e_{1g}^3}$

$E = 2(\alpha + 2\beta) + 3(\alpha + \beta) = \underline{5\alpha + 7\beta}$

Metallic and Ionic Solids

15

1. (a) P (group V) has one more valence electron than Ge(group IV); therefore Ge doped with P forms an n-type semiconductor.

   (b) In (group III) has one less valence electron than Ge(group IV); therefore Ge doped with I_n forms a p-type semiconductor.

3. The valence electrons of Mg are 3s electrons, those of O are 2p electrons. We expect that the O2p electrons will be much lower in energy than the Mg3S electrons. The energy level diagram is expected to look like the figure below.

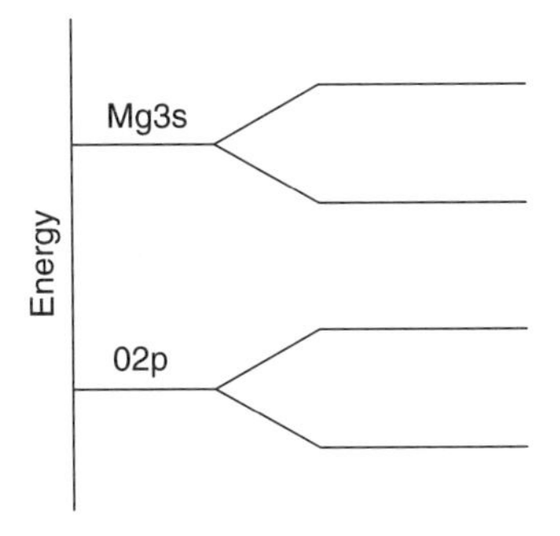

   There are two degenerate O2p orbitals available for band formation, which results in a doubly degenerate O2p band in oxygen. These bands can hold 4*N* electrons, 2*N* of which are contributed by Mg. Only the lower band is occupied and because of the big band gap MgO is an insulator. Effectively electrons have been transferred from Mg to O as in the ionic model.

5. We need to find the enthalpy change for

   $CaCl_2(s) \rightarrow Ca^{2+}(g) + 2Cl^-(g) \qquad \Delta H^\ominus_L$

   We set up a Born-Haber cycle.

(1) Sublimation of Ca(s)	+178.2 kJ mol^{-1}
(2) Ionization of Ca(g) to Ca^{2+}(g)	+1740 kJ mol^{-1}
(3) Dissociation of Cl_2(g)	+242 kJ mol^{-1}
(4) Electron attachment to 2Cl(g)	−698 kJ mol^{-1}

(5) Formation of $CaCl_2(s)$ $\quad -\Delta H^{\ominus}_{L}$

The sum of the above 5 steps corresponds to

$Ca(s) + Cl_2(g) \rightarrow CaCl_2(s) \qquad \Delta_f H^{\ominus} = -795.8 \text{ kJ mol}^{-1}$

Therefore,

$\Delta H^{\ominus}_{L} = (178 + 1740 + 242 - 698 + 796) \text{ kJ mol}^{-1}$

$\quad = \underline{2258 \text{ kJ mol}^{-1}}$

7. We need to evaluate the ratio of the factors that depend on d for the two compounds.

$$\frac{\Delta H^{\ominus}_{L}(\text{CaO})}{\Delta H^{\ominus}_{L}(\text{SrO})} = \frac{\frac{1}{d} \times \left(1 - \frac{d^*}{d}\right) (\text{CaO})}{\frac{1}{d} \times \left(1 - \frac{d^*}{d}\right) (\text{SrO})}$$

$$= \frac{\frac{1}{240} \times \left(1 - \frac{34.5}{240}\right)}{\frac{1}{256} \times \left(1 - \frac{34.5}{256}\right)}$$

$$= \underline{1.06}$$

9. The points and planes are shown in the figure below.

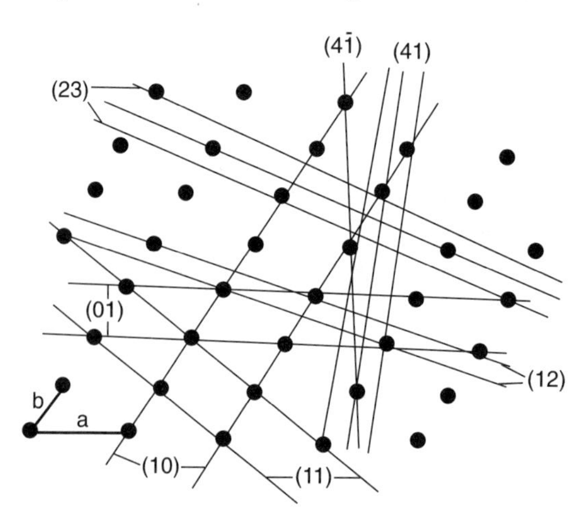

11.

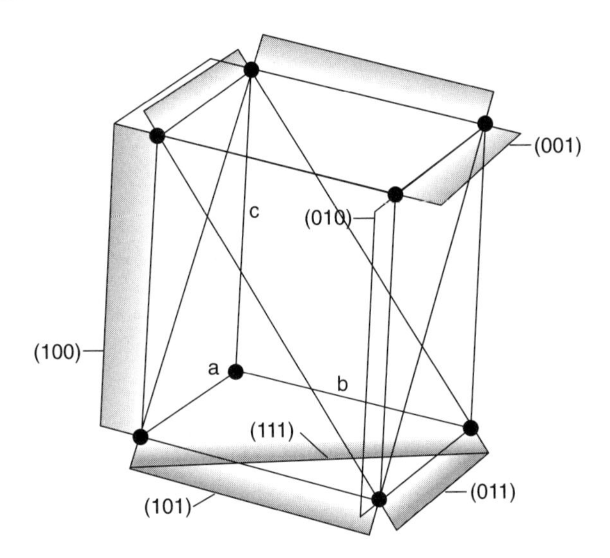

13. $\dfrac{1}{d^2} = \dfrac{h^2}{a^2} + \dfrac{k^2}{b^2} + \dfrac{l^2}{c^2}$ [15.5]

For a cubic unit cell in which $a = b = c$

$$d_{\text{hkl}} = \frac{a}{(h^2 + k^2 + l^2)^{1/2}}$$

Therefore,

$$d_{111} = \frac{a}{3^{1/2}} = \frac{532\text{ pm}}{3^{1/2}} = \underline{307\text{ pm}}$$

$$d_{211} = \frac{a}{6^{1/2}} = \frac{532\text{ pm}}{6^{1/2}} = \underline{217\text{ pm}}$$

$$d_{100} = a = \underline{532\text{ pm}}$$

15. $\lambda = 2d\sin\theta$ [15.6]

$$= 2 \times (97.3\text{ pm}) \times (\sin 19.85°) = \underline{66.1\text{ pm}}$$

17. $d_{100} = a = 350$ pm

$\rho = \dfrac{NM}{VN_\text{A}}$, implying that

$$N = \frac{\rho V N_\text{A}}{M} = \frac{(0.53 \times 10^6\text{ g m}^{-3}) \times (350 \times 10^{-12}\text{ m})^3 \times (6.022 \times 10^{23}\text{ mol}^{-1})}{6.94\text{ g mol}^{-1}}$$

$$= 1.97$$

An fcc cubic cell has $N = 4$ and a bcc unit cell has $N = 2$. Therefore, lithium has a <u>bcc unit cell</u>.

19. We use

$$\rho(x) = \frac{1}{V}\left\{F_0 + 2\sum_{h=1}^{\infty} F_h \cos(2h\pi x)\right\}$$

Because V is unknown, we work with

$$\begin{aligned} V\rho(x) = 30 &+ 16.4\cos(2\pi x) + 13.0\cos(4\pi x) + 8.2\cos(6\pi x) \\ &+ 11\cos(8\pi x) - 4.8\cos(10\pi x) + 10.8\cos(12\pi x) \\ &+ 6.4\cos(14\pi x) - 2.0\cos(16\pi x) + 2.2\cos(18\pi x) \\ &+ 13.0\cos(20\pi x) + 10.4\cos(22\pi x) - 8.6\cos(24\pi x) \\ &- 2.4\cos(26\pi x) - 0.2\cos(28\pi x) + 4.2\cos(30\pi x) \end{aligned}$$

A plot of $V\rho(x)$ is shown in the figure below.

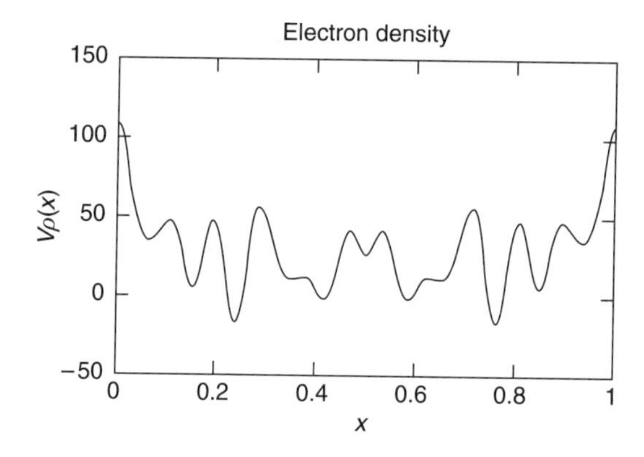

21. We need the packing fraction for hexagonal close-packing, which is 0.740. Therefore the density of the solid virus sample is

$0.740 \times 1.00 \text{ g cm}^{-3} = \underline{0.740 \text{ g cm}^{-3}}$

23. Refer to Figures 15.27 and 15.29 of the text.

(a) $\underline{12}$ nearest neighbors

(b) $\underline{6}$ next nearest neighbors

Nearest neighbors touch each other along the diagonal of a face; therefore with a = length of side of unit cell

$(2d)^2 = 2a^2$ $\quad d$ = distance between neighbors

$2d = \sqrt{2}\,a$

$d = a/\sqrt{2} = 500 \text{ nm}/\sqrt{2} = \underline{354 \text{ nm}}$

For next nearest neighbors, $d = a = 500$ nm.

25. $V = 651 \text{ pm} \times 651 \text{ pm} \times 934 \text{ pm} = \underline{3.96 \times 10^{-28} \text{ m}^3}$

If we assume a simple tetragonal unit cell, then there is one formula unit per unit cell.

$571.81 \text{ g mol}^{-1} \times (1 \text{ mol}/6.022 \times 10^{23} \text{ atoms}) = 94.3 \times 10^{-23} \text{ g atom}^{-1}$

$d = m/V = 94.93 \times 10^{-23} \text{ g}/3.96 \times 10^{-28} \text{ m}^3$

$= \underline{2.40 \times 10^6 \text{ g m}^{-3}}$

27. $$d_{hkl} = \left[\left(\frac{h}{a}\right)^2 + \left(\frac{k}{b}\right)^2 + \left(\frac{l}{c}\right)^2\right]^{-1/2}$$

(a) $$d_{321} = \left\{\left(\frac{3}{812}\right)^2 + \left(\frac{2}{947}\right)^2 + \left(\frac{1}{637}\right)^2\right\}^{-1/2} \text{pm}$$

$= \underline{220 \text{ pm}}$

(b) $d_{642} = \frac{1}{2} \text{d}_{321} = \underline{110 \text{ pm}}$

Molecular Substances

16

1. $\chi(\mathrm{H}) = 2.1$ $\chi(\mathrm{Cl}) = 3.0$ $\Delta\chi = 0.9$

We use $\mu = \Delta\chi \mathrm{D}$ [16.2]

$= \underline{0.9\ \mathrm{D}}$

$\mu = 0.9 \times 3.3 \times 10^{-30}\ \mathrm{Cm}$

$= \underline{3.0 \times 10^{-30}\ \mathrm{Cm}}$

The experimental value is 1.08 D.

3. Refer to diagram **4** of the text and eqn 16.3. Here $\mu_1 = \mu_2$; therefore

$\mu_{\mathrm{res}} = (2\mu_1^2 + 2\mu_1^2 \cos\theta)^{1/2}$

(a) ortho-xylene

$\mu_{\mathrm{res}} = (2\mu_1^2 + 2\mu_1^2 \cos 60°)^{1/2} = (3\mu_1^2)^{1/2} = \sqrt{3}\mu_1$

$= \sqrt{3} \times 0.4\ \mathrm{D} = \underline{0.7\ \mathrm{D}}$

(b) meta-xylene

$\mu_{\mathrm{res}} = (2\mu_1^2 + 2\mu_1^2 \cos 120°)^{1/2} = \mu_1 = \underline{0.4\ \mathrm{D}}$

(c) para-xylene, $\mu_{\mathrm{res}} = \underline{0}$

The <u>para-xylene</u> value is exact by symmetry.

5. $\mu = (\mu_1^2 + \mu_2^2 + 2\mu_1\mu_2 \cos\theta)^{1/2}$ [16.3]

$= \{(1.5)^2 + (0.80)^2 + 2 \times 1.5 \times 0.80 \times (\cos 109.5°)\}^{1/2}\ \mathrm{D}$

$= \underline{1.4\ \mathrm{D}}$

7. $\mu_x = 0.02e \times (-86\ \mathrm{pm}) + 0.02e \times (34\ \mathrm{pm}) + 0.06e \times (-195\ \mathrm{pm})$
$+ 0.18e \times (-199\ \mathrm{pm}) - 0.36e \times (-101\ \mathrm{pm}) + 0.45e \times (82\ \mathrm{pm})$
$- 0.38e \times (199\ \mathrm{pm}) + 0.18e \times (-80\ \mathrm{pm}) - 0.38e \times (49\ \mathrm{pm})$
$+ 0.42e \times (129\ \mathrm{pm})$

$= -29.76e\ \mathrm{pm}$

$\mu_y = 0.02e \times (118\ \mathrm{pm}) + 0.02e \times (146\ \mathrm{pm}) + 0.06e \times (70\ \mathrm{pm})$
$+ 0.18e \times (-1\ \mathrm{pm}) - 0.36e \times (-11\ \mathrm{pm}) + 0.45e \times (-15\ \mathrm{pm})$
$-0.38e \times (16\ \mathrm{pm}) + 0.18e \times (-110\ \mathrm{pm}) - 0.38e \times (-107\ \mathrm{pm})$
$+ 0.42 \times (-146\ \mathrm{pm})$

$= +21.29e\ \mathrm{pm}$

$$\mu_z = 0.02e \times (37 \text{ pm}) + 0.02e \times (-98 \text{ pm}) + 0.06e \times (-38 \text{ pm})$$
$$+ 0.18e \times (-100 \text{ pm}) - 0.36e \times (-126 \text{ pm}) + 0.45e \times (34 \text{ pm})$$
$$- 0.38e \times (-38 \text{ pm}) + 0.18e \times (-111 \text{ pm}) - 0.38e \times (88 \text{ pm})$$
$$+ 0.42e \times (126 \text{ pm})$$
$$= +53.1e \text{ pm}$$

The magnitude of the dipole moment is given by

$$\mu = (\mu_x^2 + \mu_y^2 + \mu_z^2)^{1/2} \qquad [16.4a]$$
$$= [(-29.76)^2 + (21.29)^2 + (53.1)^2]^{1/2}\text{e pm}$$
$$= 64.5\text{e pm}$$
$$= 64.5 \times 1.602 \times 10^{-19} \text{ C} \times 10^{-12} \text{ m}$$
$$= 1.03 \times 10^{-29} \text{ Cm}$$
$$= \underline{3.10 \text{ D}}$$

9. We assume that the dipole of the water molecule and the Li^+ ion are collinear and that the separation of charges in the dipole is smaller than the distance to the ion. This is certainly justified at 300 pm, but may not be at 100 pm. With these assumptions we can use eqn 16.5(a) of the text. To flip the water molecule over requires twice the energy of interaction given by eqn 16.5(a)

$$E = 2V = \frac{2q_2\mu_1}{4\pi\varepsilon_0 r^2} \qquad |q_2| = e = 1.602 \times 10^{-19} \text{ C}$$

$$= \frac{2 \times 1.602 \times 10^{-19} \text{ C} \times 1.85 \times 3.336 \times 10^{-30} \text{ Cm}}{4\pi \times 8.854 \times 10^{-12} \text{ J}^{-1} \text{ C}^2 \text{ m}^{-1} \times r^2}$$

$$= \frac{1.777 \times 10^{-38} \text{ J m}^2}{r^2}$$

(a) $$E = \frac{1.777 \times 10^{-38} \text{ J m}^2}{(1.00 \times 10^{-10} \text{ m})^2} = 1.777 \times 10^{-18} \text{ J}$$

molar energy = $N_A \times E = \underline{1070 \text{ kJ mol}^{-1}}$

(b) $$E = \frac{1.777 \times 10^{-38} \text{ J m}^2}{(3.00 \times 10^{-10} \text{ m})^2} = 1.975 \times 10^{-19} \text{ J}$$

molar energy = $N_A \times E = \underline{119 \text{ kJ mol}^{-1}}$

11. (a) The kinetic energy per mole is given by

$E = 3/2\ RT = 3.7\ \text{kJ mol}^{-1}$ at 298 K

(b) For a mole of molecules in a 10 L volume, the volume occupied per molecule is on average

$$v = \frac{10 \times 10^{-3}\ \text{m}^3\ \text{mol}^{-1}}{6.02 \times 10^{23}\ \text{mol}^{-1}} = 1.66 \times 10^{-26}\ \text{m}^3$$

This places the molecules at an average distance of $r = v^{1/3}$ with respect to each other.

$r = (1.66 \times 10^{-26}\ \text{m}^3)^{1/3} = 2.55 \times 10^{-9}\ \text{m} = 2.55\ \text{mm}$

Then per pair of molecules

$$V = -\frac{2 \times (1.08\ \text{D})^4 \times (3.336 \times 10^{-30}\ \text{Cm})^4}{3 \times (4\pi \times 8.854 \times 10^{-12}\ \text{J}^{-1}\ \text{C}^2\ \text{m}^{-1})^2 \times 1.38 \times 10^{-23}\ \text{JK}^{-1} \times 298\ \text{K} \times (2.55 \times 10^{-9}\ \text{m})^6}$$

$$= -8.03 \times 10^{-27}\ \text{J}$$

Each molecule has an average of 6 nearest neighbors, but we count a pair interaction only once. Then the total potential energy per mole in this sample is

$V = -3 \times N_A \times 8.03 \times 10^{-27}\ \text{J} = \underline{-0.014\ \text{J}}$

This potential energy is exceedingly small compared to 3.7 kJ mol^{-1}, so the kinetic theory of gases is justifiable for this sample.

13. $V = -\dfrac{\mu_1^2 \alpha_2'}{4\pi \varepsilon_0 r^6}$ [16.10]

$$= -\frac{(1.26 \times 3.336 \times 10^{-30}\ \text{Cm})^2 \times 10.4 \times 10^{-30}\ \text{m}^3}{(4\pi \times 8.854 \times 10^{-12}\ \text{J}^{-1}\ \text{C}^2\ \text{m}^{-1}) \times (4.0 \times 10^{-9}\ \text{m})^6}$$

$$= -4.03 \times 10^{-28}\ \text{J}$$

$$= \underline{-2.4 \times 10^{-4}\ \text{J mol}^{-1}}$$

This value seems exceedingly small. The distance suggested in the exercise may be too large compared to typical values.

15. The geometry shown is linear, that is, θ in structure (11) is 0°. The partial charges are as given in Table 16.1 and Example 16.2. Distances in structure (21) are not given, but we may take r(O – H) to be 95.7 pm and R (O .... N) to be 200 pm as a typical value.

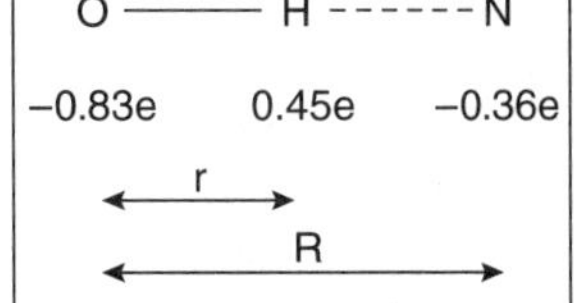

Then

$$V_{ON} = \frac{N_A Z_0 Z_N e^2}{4\pi\varepsilon_0 R} \qquad \text{[Example 16.2]}$$

and

$$V_{NH} = \frac{N_A Z_N Z_H e^2}{4\pi\varepsilon_0 (R - r)} \qquad \text{[Example 16.2]}$$

$$\frac{N_A e^2}{4\pi\varepsilon_0} = 1.389 \times 10^{-4} \text{ J m}$$

$$V = V_{ON} + V_{NH} = 1.389 \times 10^{-4} \text{ Jm} \left[\frac{Z_O Z_N}{R} + \frac{Z_N Z_H}{(R-r)} \right]$$

$$= 1.389 \times 10^{-4} \text{ J m} \left[\frac{0.83 \times 0.36}{2.00 \times 10^{-10} \text{ m}} - \frac{0.36 \times 0.45}{(2.00 - 0.957) \times 10^{-10} \text{ m}} \right]$$

$$= 1.389 \times 10^{-4} \text{ J m} \times (0.1494 - 0.1553) \times 10^{10} \text{ m}$$

$$= \underline{-8.2 \text{ kJ mol}^{-1}}$$

17. In the dimer, the dipole moments will tend to cancel, at least partially, depending on the exact orientation. As the temperature increases collisions and internal thermal agitation will break up the dimers and the dipole moments will not cancel.

19. The potential energy has the form

$$V = 4\varepsilon \left[A e^{-r/\sigma} - \left(\frac{\sigma}{r} \right)^6 \right]$$

and is sketched in the figure below.

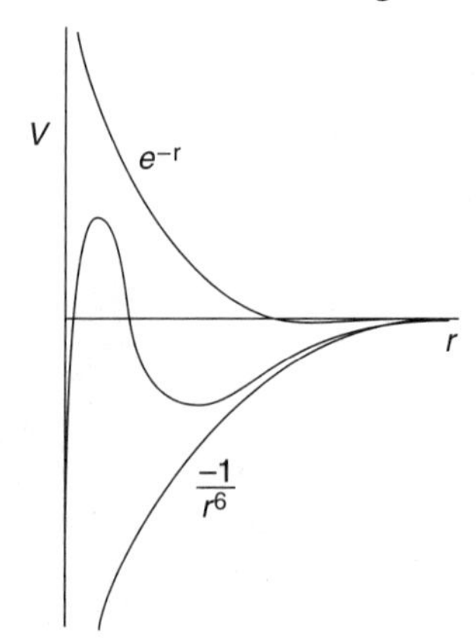

The minimum occurs where

$$\frac{dV}{dr} = 4\varepsilon\left(\frac{-A}{\sigma}\,e^{-r/\sigma} + \frac{6\sigma^6}{r^7}\right) = 0$$

which occurs at the solution of

$$\frac{\sigma^7}{r^7} = \frac{A}{6}\,e^{-r/\sigma}$$

Solve this equation numerically. As an example, when $A = \sigma = 1$, a minimum occurs at $r = \underline{1.63}$.

21. In both cases, we view the liquid as a collection of hard spheres, then the radial distribution function will consist of a series of sharp spikes at distances of the nearest neighbors, next nearest neighbors, and so on.

(a) For a cubic close-packed arrangement, the maxima will occur at $\dfrac{a}{\sqrt{2}}$ and a, where a is the length of the unit cell ($a = 2\sqrt{2}r$). The first minimum occurs at

$$\frac{a + \sqrt{2}a}{2\sqrt{2}}.$$

(b) For a body-centered cubic, the maxima occur at $\dfrac{\sqrt{3}}{2}a$ and a. The minima will occur halfway between.

These functions are sketched in the figures below.

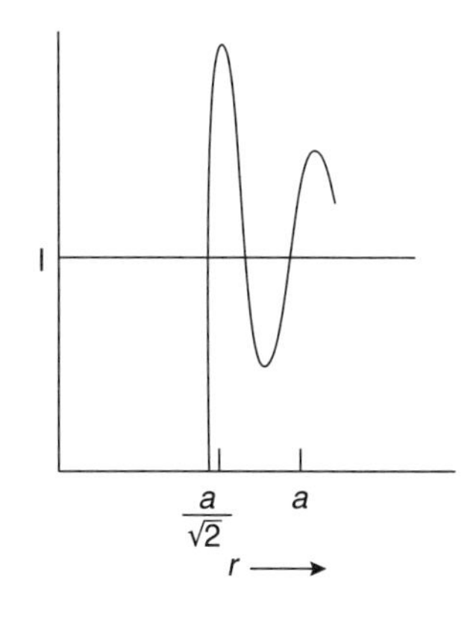

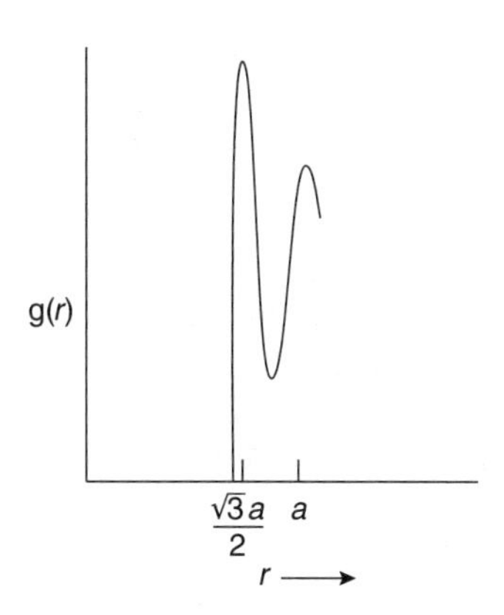

23. $d = (2Dt)^{1/2}$ [16.18]

Solve for t,

$$t = \frac{d^2}{2D}$$

(a) $t = \dfrac{(1 \times 10^{-3})^2}{2 \times 5.22 \times 10^{-10}\ \text{m}^2\ \text{s}^{-1}} = \underline{1 \times 10^{-3}\ \text{s}}$

(b) $\underline{1 \times 10^5 \text{ s}}$

(c) $\underline{1 \times 10^9 \text{ s}}$

25. We use as in Exercise 16.23 the relation

$$t = \frac{d^2}{2D}$$

$$\text{with } D = \frac{kT}{6\pi\eta a} = \frac{1.38 \times 10^{-23} \text{ J K}^{-1} \times 310 \text{ K}}{6\pi \times 0.010 \text{ kg m}^{-1} \text{ s}^{-1} \times 2.00 \times 10^{-10} \text{ m}}$$

$$= 1.14 \times 10^{-10} \text{ m}^2 \text{ s}^{-1}$$

$$t = \frac{(0.50 \times 10^{-9} \text{ m})^2}{2 \times 1.14 \times 10^{-10} \text{ m}^2 \text{ s}^{-1}} = \underline{1.1 \times 10^{-9} \text{ s}}$$

27.

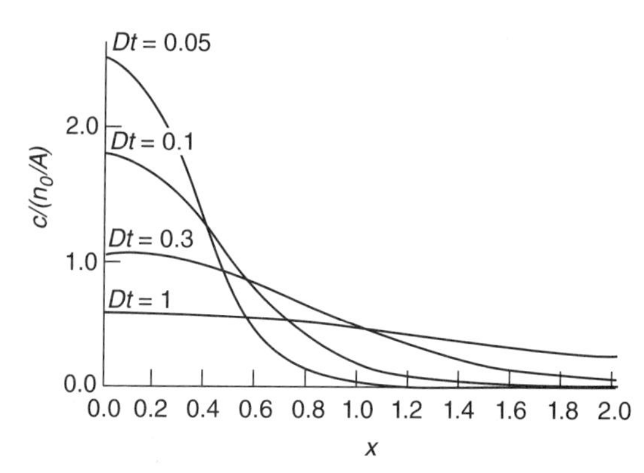

Concentration, c, as a function of time and position. n_0 is the number of moles of toxin initially confined to 5 mm at the end of the tube. A is the cross-sectional area of the tube. The units of Dt and x are arbitrary, but related so that Dt/x^2 is dimensionless. For example, if x is in cm, Dt would be in cm^2, and so for $D = 10^{-5}$ cm^2 s^{-1}, "$Dt = 0.1$" corresponds to $t = 10^4$s.

29. At high pH all proteins will be negatively charged because the carboxyls are negative (COO^-) and the amino acids are neutral (NH_2). If we start the sample on the high pH side, the negatively charged proteins will move toward the positive electrode until each one reaches its isoelectric point (pH) and stops moving.

Molecular Rotations and Vibrations 17

1. (a) $\nu = \dfrac{c}{\lambda} = \dfrac{2.998 \times 10^{8}\ \text{m s}^{-1}}{6.70 \times 10^{-7}\ \text{m}} = \underline{4.47 \times 10^{14}\ \text{s}^{-1}}$

(b) $\tilde{\nu} = \dfrac{1}{\lambda} = \dfrac{1}{6.70 \times 10^{-7}} = 1.49 \times 10^{6}\ \text{m}^{-1} = \underline{1.49 \times 10^{4}\ \text{cm}^{-1}}$

3. 55 m.p.h. = $24.6\ \text{m s}^{-1}$

we use $\nu' = \left(\dfrac{1 + s/c}{1 - s/c}\right)^{1/2} \nu$ [17.36]

$$\text{or } \lambda' = \left(\frac{1 - s/c}{1 + s/c}\right)^{1/2} \lambda$$

$$= \left(\frac{1 - 24.6/2.998 \times 10^{8}}{1 + 24.6/2.998 \times 10^{8}}\right)^{1/2} \lambda$$

$$= \underline{0.999\,999\,918 \times 660\ \text{nm}}$$

The formula for λ' can be rewritten after expanding the numerator and denominator

$$(1 \pm x)^{1/2} = 1 \pm \frac{x}{2} + \ldots$$

Then

$$\lambda' \approx (1 - s/c)\lambda$$

Now solve for s such that $\lambda' = 520$ nm

$$s = \left(1 - \frac{\lambda'}{\lambda}\right) c$$

$$= \left(1 - \frac{520}{660}\right) \times 2.998 \times 10^{8}\ \text{m s}^{-1}$$

$$= \underline{6.36 \times 10^{7}\ \text{m s}^{-1}} = 1.4 \times 10^{8}\ \text{mph}$$

5. $\delta\tilde{\nu} = \dfrac{5.3\ \text{cm}^{-1}}{\tau/\text{ps}}$ implying that $\tau = \dfrac{5.3\ \text{ps}}{\delta\tilde{\nu}/\text{cm}^{-1}}$

(a) $\tau = \dfrac{5.3\ \text{ps}}{0.1} = \underline{53\ \text{ps}}$

(b) $\tau = \dfrac{5.3\ \text{ps}}{1} = \underline{5\ \text{ps}}$

(c) $\lambda = \dfrac{2.998 \times 10^{8}\ \text{m s}^{-1}}{1.0 \times 10^{9}\ \text{s}^{-1}} = 0.300\ \text{m} = 30.0\ \text{cm}$

$\tilde{\nu} = \dfrac{1}{\lambda} = 0.0333\ \text{cm}^{-1}$

$\tau = \dfrac{5.3\ \text{ps}}{0.0333} = \underline{1.6 \times 10^{2}\ \text{ps}}$

7. $E_{\text{J}} = hBJ(J+1) \approx hBJ^{2}$ for large J

$J = \left(\dfrac{E_{\text{J}}}{hB}\right)^{1/2} \qquad B = \dfrac{\hbar}{4\pi \text{I}} \qquad I = mR^{2}$

$J = \dfrac{\pi R}{h}\left(8mE_{\text{J}}\right)^{1/2}$

$= \dfrac{\pi \times 0.70\ \text{m}}{6.63 \times 10^{-34}\ \text{J s}} \times (8 \times 0.75\ \text{kg} \times 0.2\ \text{J})^{1/2}$

$\approx \underline{4 \times 10^{33}}$

9. $B = \dfrac{\hbar}{4\pi I} = \dfrac{\text{J s}}{\text{kg m}^{2}} = \dfrac{\text{kg m}^{2}\ \text{s}^{-2}\ \text{s}}{\text{kg m}^{2}} = \text{s}^{-1}$

So we see that the unit of B is $\text{s}^{-1} = \text{Hz}$

(a) $B = \dfrac{1.0546 \times 10^{-34}\ \text{J s}}{4\pi \times 4.599 \times 10^{-48}\ \text{kg m}^{2}} = \underline{1.825 \times 10^{12}\ \text{Hz}}$

(b) $B = \dfrac{1.0546 \times 10^{-34}\ \text{J s}}{4\pi \times 9.194 \times 10^{-48}\ \text{kg m}^{2}} = \underline{9.128 \times 10^{11}\ \text{Hz}}$

(c) $B = \dfrac{1.0546 \times 10^{-34}\ \text{J s}}{4\pi \times 6.67 \times 10^{-46}\ \text{kg m}^{2}} = \underline{1.26 \times 10^{10}\ \text{Hz}}$

(d) $B = \underline{1.26 \times 10^{10}\ \text{Hz}}$

11.

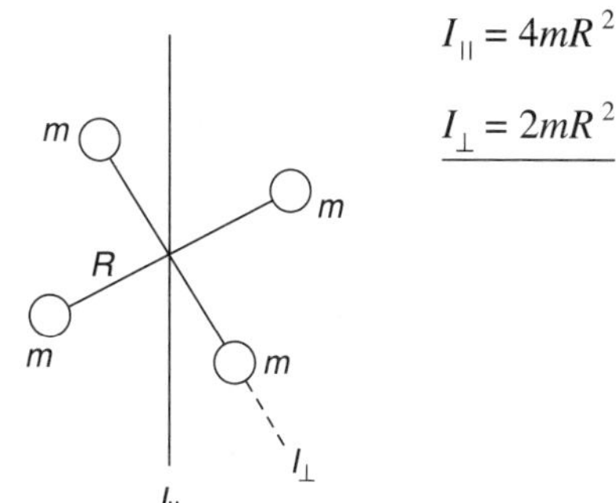

$I_{||} = 4mR^2$

$I_{\perp} = 2mR^2$

13. Polar molecules show a pure rotational spectrum. Therefore, select the polar molecules based on their well-known structures.

(a), (b), (c), and (d) have dipole moments and will have a pure rotational spectrum. (e) will not.

15. Methane is a spherical rotor; hence $I_{||} = I_{\perp}$ and $A = B$ and

$E_{J,K} = hcBJ(J + 1)$

and there is one energy level with $J = 10$.

Comment: There is a degeneracy of $2J + 1 = 21$ associated with this level, so there are a total of 21 quantum states.

17. $E_J = hBJ(J + 1)$

$$\Delta E = E_{J+1} - E_J = hB(J + 1)(J + 2) - hBJ(J + 1)$$
$$= 2hB(J + 1)$$

The separations of the lines are therefore

$2hB, 4hB, 6hB,$

(a) The frequencies of the lines are

$2B, 4B, 6B, \ldots$

$= 636\text{ GHz}, 1272\text{ GHz}, 1908\text{ GHz} \ldots$

(b) The wavenumbers of the lines are given by

$$\tilde{\nu} = \frac{\nu}{c} = \frac{636 \times 10^9\text{ s}^{-1}}{2.998 \times 10^8\text{ m s}^{-1}} = 2121\text{ m}^{-1} = 21.21\text{ cm}^{-1}$$

So the set of lines corresponds to

$21.21\text{ cm}^{-1}, 42.42\text{ cm}^{-1}, 63.63\text{ cm}^{-1}, \ldots$

19. $B = \dfrac{\hbar}{4\pi I} = 11.70\ \text{GHz} = 11.70 \times 10^9\ \text{s}^{-1}$

$I = 2m_0R^2$

Solve for R^2 in terms of B

$$R^2 = \frac{\hbar}{8\pi m_0 B}$$

$$R^2 = \frac{1.0546 \times 10^{-34}\ \text{J s}}{8\pi \times 15.995 \times 1.66054 \times 10^{-27}\ \text{kg} \times 11.70 \times 10^9\ \text{s}^{-}}$$

$$= 1.350 \times 10^{-20}\ \text{m}^2$$

$R = \underline{1.162 \times 10^{-10}\ \text{m} = 116.2\ \text{pm}}$

21. $\nu = \dfrac{1}{2\pi}\left(\dfrac{k}{\mu}\right)^{1/2}$ [17.14b]

(a) $\mu = \dfrac{m_\text{C}m_\text{O}}{m_\text{C} + m_\text{O}} = \dfrac{12.0000 \times 15.9949}{12.0000 + 15.9949} \times 1.66054 \times 10^{-27}\ \text{kg}$

$$= 1.1385 \times 10^{-26}\ \text{kg}$$

$$\nu = \frac{1}{2\pi}\left(\quad 908\ \text{Nm}^{-1}\quad\right)^{1/2}$$

$= 4.49 \times 10^{13}\ \text{s}^{-1} = \underline{4.49 \times 10^{13}\ \text{Hz}}$

(b) $\mu = 1.1910 \times 10^{-26}\ \text{kg}$

$\nu = 4.39 \times 10^{13}\ \text{s}^{-1} = \underline{4.39 \times 10^{13}\ \text{Hz}}$

23. As shown in the solution to Exercise 17.22

$k = (2\pi c\tilde{\nu})^2\mu$

$$\mu(\text{HF}) = \frac{1.0078 \times 18.9908}{1.0078 + 18.9908}\ \text{u} = 0.9570\ \text{u}$$

$$\mu(\text{H}^{35}\text{Cl}) = \frac{1.0078 \times 34.9688}{1.0078 + 34.9688}\ \text{u} = 0.9796\ \text{u}$$

$$\mu(\text{H}^{81}\text{Br}) = \frac{1.0078 \times 80.9163}{1.0078 + 80.9163}\ \text{u} = 0.9954\ \text{u}$$

$$\mu(\text{H}^{127}\text{I}) = \frac{1.0078 \times 126.9045}{1.0078 + 126.9045}\ \text{u} = 0.9999\ \text{u}$$

Using the above equation draw up the following table:

	HF	HCl	HBr	HI
$\tilde{\nu}$	4141.3	2988.9	2649.7	2309.5
μ/u	0.9570	0.9796	0.9954	0.9999
$k/(\text{N m}^{-1})$	967.1	515.6	411.8	314.2

25. Select those molecules in which a vibration gives rise to a change in dipole moment. It is helpful to write down the structural formulas of the compounds. The molecules that show infrared absorption are:

(b) HCl, (c) CO_2, (d) H_2O, (e) CH_3CH_3, (f) CH_4, and (g) CH_3Cl

27. The uniform expansion is depicted in the figure below.

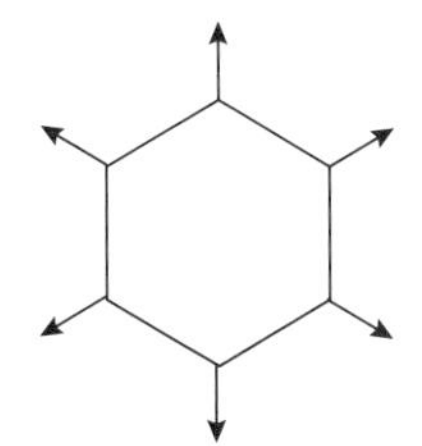

Benzene is centrosymmetric, and so the exclusion rule applies (Section 16.16).The mode is infrared inactive (symmetric breathing leaves the molecular dipole moment unchanged at zero), and therefore the mode may be Raman active (and is).

Electronic Transitions

18

1. $\log \frac{I}{I_0} = -\varepsilon[\mathrm{J}]l$ [18.1]

$= -743\ \mathrm{mol^{-1}\ L\ cm^{-1}} \times (3.25 \times 10^{-3}\ \mathrm{mol\ L^{-1}}) \times 0.25\ \mathrm{cm}$

$= -0.604$

$\frac{I}{I_0} = 0.249$

The reduction is <u>75.1%</u>

3. $\log T = -\varepsilon[\mathrm{J}]l$

Solve for [J],

$$[\mathrm{J}] = -\frac{1}{\varepsilon l}\log T$$

$$= \frac{-\log 0.602}{291\ \mathrm{L\ mol^{-1}\ cm^{-1}} \times 0.65\ \mathrm{cm}} = \underline{1.16 \times 10^{3}\ \mathrm{mol\ L^{-1}}}$$

5. $T_{\mathrm{k}} = T_{\mathrm{u}}$

k = solution of known concentration

u = solution of unknown concentration

$\log T_{\mathrm{k}} = \log T_{\mathrm{u}}$

$-\varepsilon[\mathrm{k}]l_{\mathrm{k}} = -\varepsilon[\mathrm{u}]l_{\mathrm{u}}$

Solve for [u],

$$[\mathrm{u}] = [\mathrm{k}]\left(\frac{l_{\mathrm{k}}}{l_{\mathrm{u}}}\right) = 25\ \mu\mathrm{g\ L^{-1}} \times \left(\frac{1.55\ \mathrm{cm}}{1.18\ \mathrm{cm}}\right) = \underline{33\ \mu\mathrm{g\ L^{-1}}}$$

7. $\varepsilon = -\dfrac{1}{[\mathrm{J}]l}\log\dfrac{I}{I_0}$ with $l = 0.20$ cm

We use this formula to draw up the following table.

$[Br_2]/\mathrm{mol\ L^{-1}}$	0.0010	0.0050	0.0100	0.0500	
I/I_0	0.814	0.356	0.127	3.0×10^{-5}	
$\varepsilon/(\mathrm{L\ mol^{-1}\ cm^{-1}})$	447	449	448	452	mean: $44\overline{9}$

Hence, the molar absorption coefficient is $\varepsilon = \underline{450\ \mathrm{L\ mol^{-1}\ cm^{-1}}}$

9. $l = \dfrac{-1}{\varepsilon[\mathrm{J}]}\log\dfrac{I}{I_0}$

For water, $[H_2O] \approx \dfrac{1.00\ \mathrm{kg/L}}{18.02\ \mathrm{g\ mol^{-1}}} = 55.5\ \mathrm{mol\ L^{-1}}$

and $\varepsilon[\mathrm{J}] = (55.5\ \mathrm{M}) \times (6.2 \times 10^{-5}\ \mathrm{M^{-1}\ cm^{-1}}) = 3.4 \times 10^{-3}\ \mathrm{cm^{-1}} = 0.34\ \mathrm{m^{-1}}$,

so $\dfrac{1}{\varepsilon[\mathrm{J}]} = 2.9$ m

Hence, $l/\mathrm{m} = -2.9 \times \log\dfrac{I}{I_0}$

(a) $\dfrac{I}{I_0} = 0.5$, $l = -2.9\ \mathrm{m} \times \log 0.5 = \underline{0.9\ \mathrm{m}}$

(b) $\dfrac{I}{I_0} = 0.1$, $l = -2.9\ \mathrm{m} \times \log 0.1 = \underline{3\ \mathrm{m}}$

11. $A = c_A\varepsilon_A l + c_B\varepsilon_B l = (c_A + c_B)\varepsilon^\circ l$ $\qquad \varepsilon_A = \varepsilon_B$

$A' = c_A\varepsilon_A' l + c_B\varepsilon_B' l$

We need to solve these two simultaneous equations in terms of ε°, ε_A', ε_B', l, A, and A'

all of which are assumed to be known. The result is

$$\underline{c_A = \frac{\varepsilon^\circ A' - \varepsilon_B A}{(\varepsilon_A - \varepsilon_B)\varepsilon^\circ l}}$$

$$\underline{c_B = \frac{\varepsilon^\circ \mathrm{A} - \varepsilon_B \mathrm{A}'}{(\varepsilon_A - \varepsilon_B)\varepsilon^\circ l}}$$

13. There are three isosbestic wavelengths (or wavenumbers). The presence of two or more isosbestic points is good evidence that only two solutes in equilibrium with each other are present. The solutes here being $Her(CNS)_8$ and $Her(OH)_8$.

15. Tryptophan (Trp) and tyrosine (Tyr) show the characteristic absorption of a phenyl group at about 280 nm. Cysteine (Cys) and glycine (Gly) lack the phenyl group as is evident from their spectra.

17. $E = eV = 1.602 \times 10^{-19}\ \text{C} \times 5.0\ \text{V} = \underline{8.0 \times 10^{-19}\ \text{J}}$ (or 5.0 eV)

19. $I = h\nu - \text{KE}$ [18.13]

$= 21.21\ \text{eV} - \text{KE}$

$I = 10.20$ eV, 12.98 eV, and 15.99 eV

A molecular orbital configuration such as $(1\pi)^4(3\sigma)^2(2\pi^*)^1$ could possibly account for the spectrum. The energy level diagram would like like the figure below.

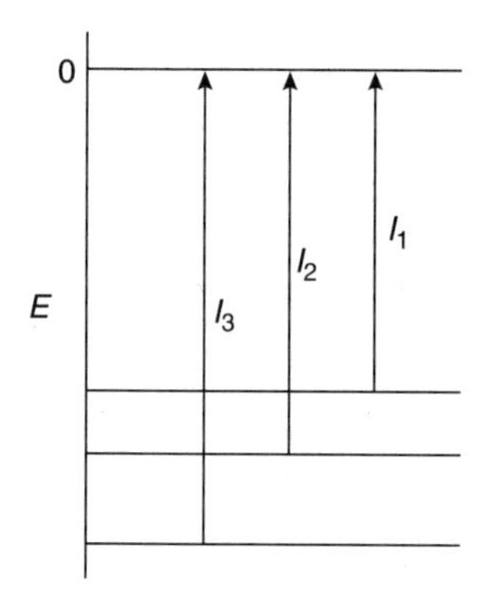

Magnetic Resonance

19

1. $E_{m_I} = -\gamma_N \hbar \mathcal{B} m_I = -g_I \mu_N \mathcal{B} m_I$ [19.1, 19.2]

   $m_I = 3/2, 1/2, -1/2, -3/2$

   $E_{m_I} = 0.4289 \times (5.051 \times 10^{-27}\ \text{J T}^{-1}) \times (7.500\ \text{T}) \times m_I$

   $= \underline{-1.625 \times 10^{-26}\ \text{J} \times m_I}$

3. $\gamma_N \hbar = g_I \mu_N$ [19.2]

   Therefore,

   $$g_I = \frac{\gamma_N \hbar}{\mu_N} = \frac{1.0840 \times 10^8\ \text{T}^{-1}\ \text{s}^{-1} \times 1.05457 \times 10^{-34}\ \text{J s}}{5.051 \times 10^{-27}\ \text{J T}^{-1}}$$

   $= \underline{2.263}$

5. $$\nu = \frac{\gamma_N \mathcal{B}}{2\pi}\ [19.4] = \frac{g_I \mu_N}{h} \mathcal{B}$$

   $$\nu = \frac{0.4036 \times 5.051 \times 10^{-27}\ \text{J T}^{-1} \times 15.00\ \text{T}}{6.626 \times 10^{-34}\ \text{J s}}$$

   $= 4.615 \times 10^7\ \text{s}^{-1} = \underline{46.15\ \text{MHz}}$

7. We use relations similar to Eqns 19.2 and 19.4. These lead to [see Exercise 19.5]

   $$\nu = \frac{g_e \mu_B \mathcal{B}}{h}$$

   where $g_e = 2.0023$ and μ_B is the Bohr magneton and is the analog of the nuclear magneton.

   $$\nu = \frac{2.0023 \times 9.274 \times 10^{-24}\ \text{J T}^{-1} \times 0.330\ \text{T}}{6.626 \times 10^{-34}\ \text{J s}}$$

   $= 9.248 \times 10^9\ \text{s}^{-1} = \underline{9.248\ \text{GHz}}$

   EPR employs microwave radiation, rather than the radio frequency radiation of NMR.

9. $\mathcal{B}_{\text{loc}} = (1 - \sigma)\mathcal{B}$

$$|\Delta \mathcal{B}_{\text{loc}}| = |(\Delta\sigma)|\, \mathcal{B} \approx |[\delta(\text{CH}_3) - \delta(\text{CHO})]|\, \mathcal{B} \qquad \left[|\Delta\sigma| \approx \left|\frac{\nu - \nu_0}{\nu_0}\right|\right]$$

$$= |(2.20 - 9.80)| \times 10^{-6}\, \mathcal{B} = 7.60 \times 10^{-6}\, \mathcal{B}$$

(a) $\mathcal{B} = 1.5\text{ T},\ |\Delta \mathcal{B}_{\text{loc}}| = 7.60 \times 10^{-6} \times 1.5\text{ T} = \underline{11\ \mu\text{T}}$

(b) $\mathcal{B} = 6.0\text{ T}\ |\Delta \mathcal{B}_{\text{loc}}| = 7.60 \times 10^{-6} \times 6.0\text{ T} = \underline{46\ \mu\text{T}}$

11. For identical nuclei with spin 1/2, there will be $N + 1$ lines from the splitting. In this case 8 lines. The lines will have relative intensities of 1:7:21:35:35:21:7:1.

These relative intensities can be determined by extending Pascals' triangle shown in (1) of the text three more rows to $N + 1 = 8$. Alternatively the intensities can also be determined from the coefficients in the expansion of $(1 + x)^N$.

13. $E = -\gamma_N\hbar(1 - \sigma_A)\, \mathcal{B} m_A - \gamma_N\hbar(1 - \sigma_X)\, \mathcal{B} m_{X_1} - \gamma_N\hbar(1 - \sigma_X)\, \mathcal{B} m_{X_2}$

As m_{X_1} and m_{X_2} can each be $\pm\frac{1}{2}$, there are a total of 6 energy levels, two of which are two-fold degenerate, for a total of eight levels. These are shown on the left of the figure below. The allowed transitions are indicated by arrows. There are 7 transitions, but only 2 transition frequencies. This follows from the selection rule for magnetic resonance transitions, which is $\Delta(m_1 + m_2) = \pm 1$. The shorter arrows represent the X transitions, the larger arrows the A transitions. Spin-spin splitting perturbs these levels as follows:

$E_{\text{spin-spin}} = hJm_A m_{X_1} + hJm_A m_{X_2}$

	$\alpha\alpha_1\alpha_2$	$\alpha\alpha_1\beta_2$	$\alpha\beta_1\alpha_2$	$\alpha\beta_1\beta_2$
$E_{\text{spin-spin}}$	$\frac{1}{2}hJ$	0	0	$-\frac{1}{2}hJ$
	$\beta\beta_1\beta_2$	$\beta\alpha_1\beta_2$	$\beta\beta_1\alpha_2$	$\beta\alpha_1\alpha_2$

There are again a total of 6 energy levels (two of which are two-fold degenerate), but they are perturbed by the amounts in the above chart. The perturbed levels are shown on the right in the figure at the top of page 127. The frequencies of the X transitions are changed by $\pm\frac{1}{2}J$, the frequencies of the A transitions by $-J, 0, +J$. A stick diagram representing the spectrum is shown in the second figure on page 127.

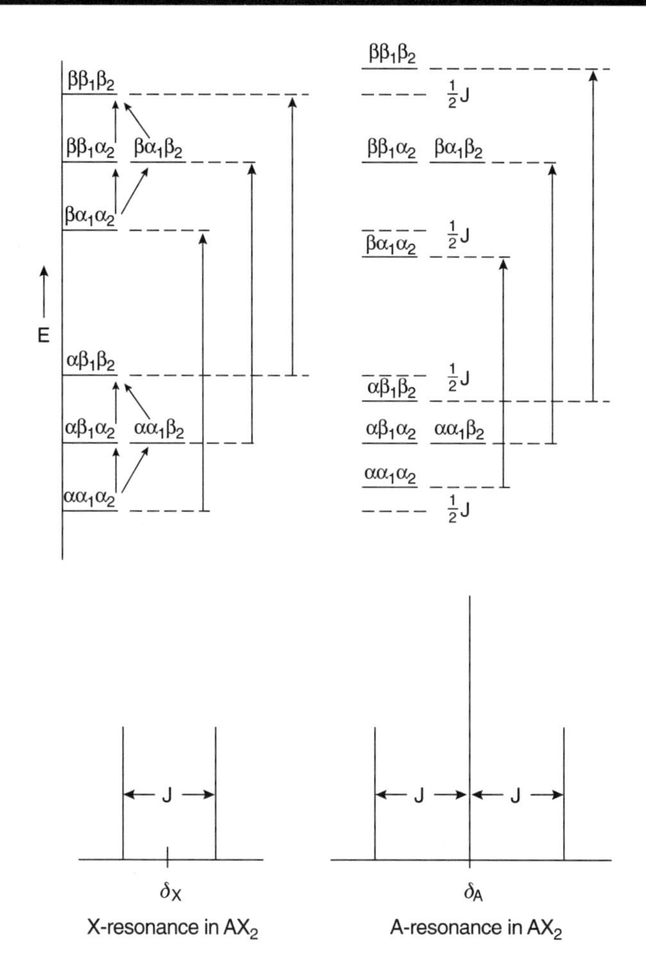

15. The four equivalent ^{19}F nuclei ($I = \frac{1}{2}$) give a single line. However, the ^{10}B nucleus ($I = 3$, 19.6 percent abundant) splits this line into $2 \times 3 + 1 = 7$ lines and the ^{11}B nucleus ($I = \frac{3}{2}$, 80.4 percent abundant) into $2 \times \frac{3}{2} + 1 = 4$ lines. The splitting arising from the ^{11}B nucleus will be larger than that arising from the ^{10}B nucleus (because its magnetic moment is larger, by a factor of 1.5). Moreover, the total intensity of the four lines due to the ^{11}B nuclei will be greater (by a factor of $80.4/19.6 \approx 4$) than the total intensity of the seven lines due to the ^{10}B nuclei. The individual line intensities will be in the ratio $\frac{7}{4} \times 4 = 7$ ($\frac{4}{7}$ the number of lines and about four times as abundant). The spectrum is sketched in the figure below.

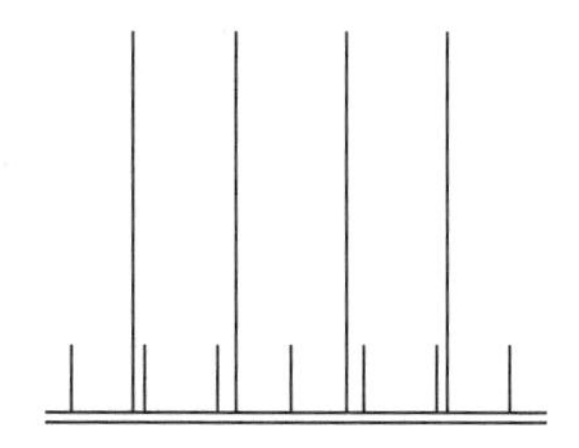

17. $^{3}J_{\text{HH}} = A + B\cos\phi + C\cos 2\phi$ [19.9]

$$\frac{\text{d}}{\text{d}\phi}\left(^{3}J_{\text{HH}}\right) = -B\sin\phi - 2\,C\sin 2\phi = 0$$

This equation has a number of solutions:

$$\phi = 0,\ \phi = n\pi,\ \phi = \pi - \arccos\left(\frac{B}{4C}\right) = \arccos\left(\frac{B}{4C}\right)$$

The first two are trivial solutions.

If $\phi = \arccos\left(\frac{B}{4C}\right)$ then, $\sin\phi = \sqrt{1 - \frac{B}{16C^2}}$

$$\sin 2\phi = 2\sin\phi\cos\phi = 2\sqrt{1 - \frac{B^2}{16C^2}}\left(\frac{B}{4C}\right)$$

$$B\sin\phi + 2C\sin 2\phi = B\sqrt{1 - \frac{B^2}{16C^2}} + 4C\sqrt{1 - \frac{B^2}{16C^2}}\left(\frac{B}{4C}\right) = 0$$

So $\frac{B}{4C} = \cos\phi$ clearly satisfies the condition for an extremum.

The second derivative is

$$\frac{\text{d}^2}{\text{d}\phi^2}\left(^{3}J_{\text{HH}}\right) = -B\cos\phi - 4C\cos 2\phi = -B\cos\phi - 4C(2\cos^2\phi - 1)$$

$$= -B\left(\frac{B}{4C}\right) - 4C\left(2\frac{B^2}{16C^2} - 1\right) = -\frac{B^2}{4C} - \frac{2B}{4C} + 4C$$

This quantity is positive if

$16C^2 > 3B^2$

This is certainly true for typical values of B and C, namely $B = -1$ Hz and C = 5 Hz. Therefore the condition for a minimum is as stated, namely, $\cos\phi = B/4C$

19. Oxygen (O_2) is a paramagnetic molecule and, as a result, its presence in solution will cause strong local fluctuating magnetic fields at the positions of the nuclei. The interaction of these fluctuating local fields with the nuclei is one of the principal contributors to relaxation. Removing the oxygen eliminates this contribution. It is to be noted that the strength of the local field caused by an electronic magnetic moment is considerably greater than that caused by nuclear moments, so even a small amount of dissolved oxygen can have a major effect on the relaxation time.

Statistical Thermodynamics

20

1. We apply the method of Illustration 20.1.

$$\frac{N_{\text{stretched}}}{N_{\text{coil}}} = e^{-\Delta E/RT}$$

$$= e^{-\frac{2.4\times 10^{3}\ \text{J mol}^{-1}}{(8.3145\ \text{J K}^{-1}\ \text{mol}^{-1})\times(293\ \text{K})}}$$

$$= \underline{0.37}$$

3. See the solution to Exercise 19.7

$$\Delta E = g_e\mu_B\mathcal{B}$$

$$\frac{N_\alpha}{N_\beta} = e^{-\Delta E/kT} = e^{-g_e\mu_B\mathcal{B}/kT}$$

$$= e^{-\frac{2.0023\times 9.274\times 10^{-24}\ \text{J T}^{-1}\times 0.33\ \text{T}}{1.38066\times 10^{-23}\ \text{J K}^{-1}\times 293\ \text{K}}}$$

$$= \underline{0.99849}$$

5. $$\frac{N_5}{N_1} = \frac{g_5}{g_1}\,e^{-28\,hB/kT} \qquad g_J = (2J+1)^2$$

$$= \frac{121}{9}\,e^{-\frac{28\times 6.62\times 10^{-34}\ \text{J s}\times 157\times 10^{9}\ \text{s}^{-1}}{1.38066\times 10^{-23}\ \text{J K}^{-1}\times 293\ \text{K}}}$$

$$= \underline{\left(\frac{121}{9}\right)0.487 = 6.55}$$

7. $$q = \Sigma_i g_i e^{-E_i/kT} = \Sigma_i g_i e^{-hc\tilde{\nu}_i/kT}$$

$$\frac{hc}{k} = \frac{6.626\times 10^{-34}\ \text{J s}\times 2.998\times 10^{8}\ \text{m s}^{-1}}{1.381\times 10^{-23}\ \text{J K}^{-1}}$$

$$= 0.014387\ \text{m K} = 1.4387\ \text{cm K}$$

$$q = 1 + 3\,\mathrm{e}^{-1.4387\text{ cm K} \times 16.4\text{ cm}^{-1}/T} + 5\mathrm{e}^{-1.4387\text{ cm K} \times 43.5\text{ cm}^{-1}/T}$$

$$= 1 + 3\mathrm{e}^{-23.59\text{ K}/T} + 5\mathrm{e}^{-62.58\text{ K}/T}$$

(a) $q = 1 + 3\mathrm{e}^{-23.59\text{ K}/10\text{ K}} + 5\mathrm{e}^{-62.58\text{ K}/10\text{ K}}$

$= \underline{1.29}$

(b) $q = 1 + 3\mathrm{e}^{-23.59\text{ K}/298\text{ K}} + 5\mathrm{e}^{-62.58\text{ K}/298\text{ K}}$

$= \underline{7.82}$

9. $q = \dfrac{1}{1 - \mathrm{e}^{-h\tilde{\nu}/kT}} = \dfrac{1}{1 - \mathrm{e}^{-h\tilde{\nu}/kT}}$

$$q = \frac{1}{1 - \mathrm{e}^{-1.4387\text{ cm K}(\tilde{\nu}/T)}}$$

$$q = \frac{1}{1 - \mathrm{e}^{-1.4387\text{ cm K} \times 560\text{ cm}^{-1}/298\text{ K}}}$$

$= \underline{1.072}$

11. (a) $q = \dfrac{kT}{\sigma hB}$ [20.6] $= \dfrac{1.381 \times 10^{-23}\text{ J K}^{-1} \times 298\text{ K}}{1 \times 6.626 \times 10^{-34}\text{ J s} \times 318 \times 10^{9}\text{ s}^{-1}}$

$= \underline{19.5}$

(b) $q = \dfrac{1.381 \times 10^{-23}\text{ J K}^{-1} \times 298\text{ K}}{2 \times 6.626 \times 10^{-34}\text{ J s} \times 11.70 \times 10^{9}\text{ s}^{-1}} = \underline{265}$

13. $q = \dfrac{1}{1 - \mathrm{e}^{-h\nu/kT}}$

$$E = \frac{NkT^2}{q}\frac{\mathrm{d}q}{\mathrm{d}T}$$

$$\frac{\mathrm{d}q}{\mathrm{d}T} = \frac{(h\nu/k)\,\mathrm{e}^{-h\nu/kT}}{(1 - \mathrm{e}^{-h\nu/kT})^2 T^2} = \frac{q^2}{T^2} = \frac{h\nu}{k}\,\mathrm{e}^{-h\nu/kT}$$

$$E = Nh\nu q\mathrm{e}^{-h\nu/kT} = \frac{Nh\nu\mathrm{e}^{-h\nu/kT}}{1 - \mathrm{e}^{-h\nu/kT}}$$

$$= \underline{\frac{Nh\nu}{\mathrm{e}^{h\nu/kT} - 1}}$$

15. (a) From the solution to Exercise 20.7 we have

$$q = 1 + 3e^{-23.59\ \mathrm{K}/T} + 5e^{-62.58\ \mathrm{K}/T}$$

$$E = \frac{NkT^2}{q}\frac{dq}{dT}$$

$$\frac{dq}{dT} = 3\left(\frac{23.59\ \mathrm{K}}{T^2}\right)e^{-23.59\ \mathrm{K}/T} + 5\left(\frac{62.58}{T^2}\right)e^{-62.58\ \mathrm{K}/T}$$

$$E = \frac{R(70.77\ \mathrm{K})e^{-23.59\ \mathrm{K}/T} + R(312.9\ \mathrm{K})e^{-62.58\ \mathrm{K}/T}}{1 + 3e^{-23.59\ \mathrm{K}/T} + 5e^{-62.58\ \mathrm{K}/T}}$$

The plot of E against T is shown in the figure below.

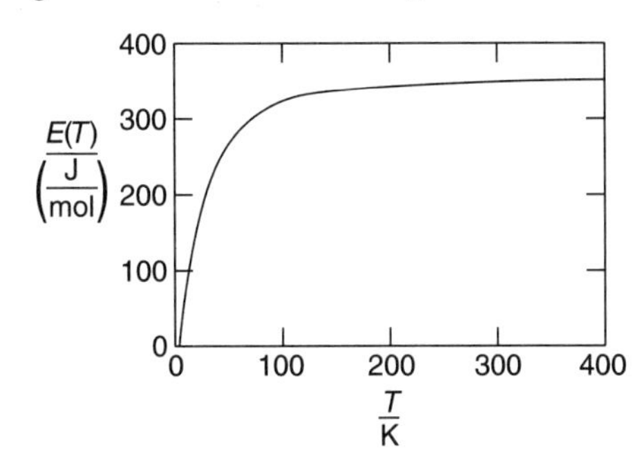

(b) Substitute $T = 298$ K into the expression for E and obtain

$E = \underline{339\ \mathrm{J\ mol^{-1}}}$

17. $S_m = k \ln 4^{N_A} = N_A k \ln 4 = R \ln 4$

$= \underline{11.5\ \mathrm{J\ K^{-1}\ mol^{-1}}}$

19. The change in entropy per micelle is

$$\Delta S = 100\ k\ln \frac{V_{\text{solution}}}{V_{\text{micelle}}}$$

In the absence of specific data on the volumes involved we can arrive at a rough value of ΔS by making some reasonable estimates. Let us assume that the micelle is spherical in shape. Let us also assume that the radius of this sphere is roughly the same as the length of the hydrocarbon chain of the amphiphile. Assume that the chain consists of 10 zig-zag carbon atoms with an average C-C-C length of 250 pm, or about 125 pm per carbon atom. Then the radius of the micelle is about 1.25 nm and the volume is

$$V_{\text{micelle}} = \frac{4}{3}\pi r^3 \approx 4r^3 = 4 \times (1.25 \times 10^{-9}\ \mathrm{m})^3$$

$$\approx 1 \times 10^{-26}\ \mathrm{m}^3$$

Assume that the volume of the solution is that of a typical 100 mL beaker.
100 mL = 10^{-4} m^3.

$$\Delta S = 100\ \text{k} \ln\left(\frac{10^{-4}\ \text{m}^3}{10^{-26}\ \text{m}^3}\right)$$

$$= 100 \times 1.38 \times 10^{-23}\ \text{J K}^{-1} \times \ln 10^{22}$$

$$= 7 \times 10^{-20}\ \text{J K}^{-1}\ \text{(per micelle)}$$

For one mole of micelles

$$\Delta S \approx N_A \times 7 \times 10^{-20}\ \text{J K}^{-1}$$

$$\approx 4 \times 10^4\ \text{J K}^{-1}\ \text{mol}^{-1}$$

$$\approx \underline{40\ \text{kJ K}^{-1}\ \text{mol}^{-1}}$$

This value seems high and is probably a result of underestimating $V_{micelle}$ for a micelle consisting of 100 amphiphiles. End effects contributing to the length of the amphiphile were neglected and the numnber of carbon atoms for amphiphiles that could form a micelle of 100 amphiphiles may be larger than the 10 assumed. Experimental evidence for the volume of micelles with 100 amphiphiles indicates typical values of about 1×10^{-25} m^3. For that volume

$$\Delta S \approx \underline{4\ \text{kJ K}^{-1}\ \text{mol}^{-1}}$$

21. $G_m^{\ominus} - G_m^{\ominus}(0) = -RT \ln \dfrac{q_m^{\ominus}}{N_A}$

In exercise 20.18 we calculated the value of $q_m^{\ominus}/N_A$ for N_2 as 3.02×10^8. Then

$$G_m^{\ominus} - G_m^{\ominus}(0) = -8.3145\ \text{J K}^{-1}\ \text{mol}^{-1} \times 298\ \text{K} \times \ln(3.02 \times 10^8)$$

$$= \underline{-48.4\ \text{kJ mol}^{-1}}$$

23. Assume $\Delta E \approx \Delta H(\text{I} - \text{I}) = 151\ \text{kJ mol}^{-1}$

$$K = \left[\frac{\left(q_{\text{I,m}}^{\ominus}\right)^2}{\left(q_{\text{I,m}}^{\ominus}\right)^2 N_A}\right] e^{-\Delta E/RT}$$

$q_{\text{I,m}}^{\ominus} = q_m^{\text{Trans}}(\text{I})q^{\text{Elec}}(\text{I})$ $\qquad$ $q^{\text{Elec}}(\text{I}) = 4$

$$\left(q_{\text{I,m}}^{\ominus}\right)^2 = q_m^{\text{Trans}}(\text{I}_2)q^{\text{Rot}}(\text{I}_2)q^{\text{Vib}}(\text{I}_2)q^{\text{Elec}}(\text{I}_2)q^{\text{Elec}}(\text{I}_2) = 1$$

See the solution to Exercise 20.18 for the partition function formulas.

$$\frac{q_{\mathrm{m}}^{\mathrm{Trans}}}{N_{\mathrm{A}}} = 2.561 \times 10^{-2}(T/\mathrm{K})^{5/2} \times (M/\mathrm{g\ mol^{-1}})^{3/2}$$

$$\frac{q_{\mathrm{m}}^{\mathrm{Trans}}(\mathrm{I_2})}{N_{\mathrm{A}}} = 2.561 \times 10^{-2} \times 500^{5/2} \times 253.8^{3/2} = 5.79 \times 10^{8}$$

$$\frac{q_{\mathrm{m}}^{\mathrm{Trans}}(\mathrm{I})}{N_{\mathrm{A}}} = 2.561 \times 10^{-2} \times 500^{5/2} \times 126.9^{3/2} = 2.05 \times 10^{8}$$

$$q^{\mathrm{Rot}}(\mathrm{I_2}) = \frac{0.6950}{\sigma} \times \frac{\mathrm{T/K}}{B/\mathrm{cm^{-1}}} = \frac{1}{2} \times 0.6950 \times \frac{500}{0.0373}$$

$$= 4.66 \times 10^{3}$$

$$q^{\mathrm{Vib}}(\mathrm{I_2}) = \frac{1}{1 - \mathrm{e}^{-a}} \qquad a = 1.4388\ \frac{\tilde{\nu}/\mathrm{cm^{-1}}}{T/\mathrm{K}}$$

$$= \frac{1}{1 - \mathrm{e}^{-1.4388 \times 214.36/500}} = 2.17$$

$$K = \frac{(2.05 \times 10^{8})^{2} \times 4^{2} \times \mathrm{e}^{-151 \times 10^{3}/8.3145 \times 500}}{5.78 \times 10^{8} \times 4.66 \times 10^{3} \times 2.17}$$

$$= \underline{1.93 \times 10^{-11}}$$

Box Exercises

Box 1.1, Exercise 1

Air is roughly 80% $N_2(g)$ and 20% $O_2(g)$. There are also some minor components, but they do not effect much the average molar mass.

$$\text{molar mass air} = 0.80 \times 28 \text{ g mol}^{-1} + 0.20 \times 32 \text{ g mol}^{-1}$$
$$\approx 29 \text{ g mol}^{-1}$$

molar mass $H_2(g) = 2.0 \text{ g mol}^{-1}$

Hence,

$$\frac{\text{density } H_2(g)}{\text{density air}} = \frac{2.0 \text{ g mol}^{-1}}{29 \text{ g mol}^{-1}} = \underline{0.069}$$

In the same volume (the volume of the balloon), the mass of air would be $29 \text{ g mol}^{-1}/2.0 \text{ g mol}^{-1} = 14.5$ times the mass hydrogen.

mass of air displaced = 14.5×10 kg = 145 kg

The payload is the difference between the mass of displaced air and the mass of the balloon (here assumed to be the mass of hydrogen).

payload = 145 kg − 10 kg = $\underline{135 \text{ kg}}$

Box 1.2, Exercise 1

We use

$$p = \frac{nRT}{V} = \frac{\rho RT}{M}$$

where M is the average molar mass of the particles. For each C^{6+} ion, there are 6 electrons. The average molar mass is therefore

$$M = \frac{12 \text{ g mol}^{-1} + 6 \times 0 \text{ g mol}^{-1}}{7} = 1.7 \text{ g mol}^{-1}$$

$$p = \frac{1.20 \times 10^3 \text{ kg m}^{-3} \times 8.3145 \text{ J K}^{-1} \text{ mol}^{-1} \times 3.5 \times 10^3 \text{ K}}{1.7 \text{ g mol}^{-1}}$$

$$= \underline{2.1 \times 10^7 \text{ Pa}}$$

Box 3.1, Exercise 1

(a) 1.0 L water ≈ 1000 g water

$$n = \frac{1000\ \text{g}}{18\ \text{g mol}^{-1}} = 56\ \text{mol}$$

$$\Delta H(\text{water}) = n\Delta_{\text{vap}}H = 56\ \text{mol} \times 40.7\ \text{kJ mol}^{-1} = 2.3 \times 10^3\ \text{kJ}$$

$$\Delta H(\text{runner}) = \underline{-2.3 \times 10^3\ \text{kJ}}$$

(b) $\Delta H = mC_p\Delta T$

$$\Delta T = \frac{\Delta H}{mC_p} = \frac{2.3 \times 10^6\ \text{J}}{6.0 \times 10^4\ \text{g} \times 4.184\ \text{J K}^{-1}\ \text{g}^{-1}}$$

$$= \underline{9.2\ \text{K}}$$

Box 4.1, Exercise 1

There is a hydrophobic effect, but it does not involve bonding of the solute molecules to water. In a sense, hydrophobic molecules are "antibonding" with respect to water. On the other hand, a hydrophilic solute may engage in hydrogen bonding with water.

Box 6.1, Exercise 1

The 97 percent saturated haemoglobin in the lungs releases oxygen in the capillary until the haemoglobin is 75 percent saturated.

100 mL of blood in the lung containing 15 g of Hb at 97 percent saturated with O_2 binds

$1.34\ \text{mL g}^{-1} \times 15\ \text{g} = 20\ \text{mL}\ O_2$

The same 100 mL of blood in the arteries would contain

$$20\ \text{mL}\ O_2 \times \frac{75\%}{97\%} = 15.5\ \text{mL}$$

Therefore, about (20 – 15.5) mL or <u>4.5 mL</u> of O_2 is given up in the capillaries to body tissue.

Box 6.2, Exercise 1

$$v = \frac{[\text{EB}]_{\text{bound}}}{[\text{M}]}$$

$$[\text{EB}]_{\text{bound}} = [\text{EB}]_{\text{in}} - [\text{EB}]_{\text{out}}$$

Draw up the following table.

$[EB]_{out}/(\mu mol\ L^{-1})$	0.042	0.092	0.204	0.526	1.150
$[EB]_{bound}/(\mu mol\ L^{-1})$	0.250	0.498	1.000	2.005	3.000
ν	0.250	0.498	1.000	2.005	3.000
$\nu/[EB]_{out}/(2\ \mu mol^{-1})$	5.95	5.41	4.90	3.81	2.61

A plot of $\nu/[EB]_{out}$ is shown in the figure below.

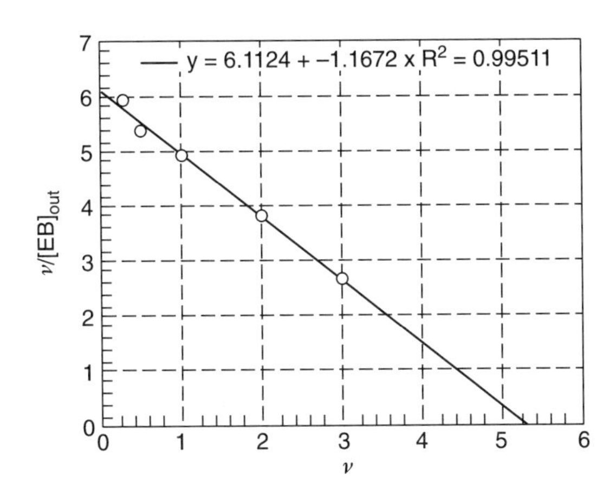

The slope is -1.167 L μmol^{-1}, hence K = <u>1.167 L μmol^{-1}</u>

The intercept at $\nu = 0$ is <u>$N = 5.24$</u> and this is the average number of binding sites per oligonucleotide.

The close fit of the data to a straight line indicates that the identical and independent sites model <u>is applicable</u>.

Box 7.1, Exercise 1

Refer to the data in Box 3.1. Carbohydrates provide about 17 kJ g^{-1} of energy. Sugar is a carbohydrate.

$$\frac{17\ \text{kJ g}^{-1}}{4.184\ \text{kJ/kcal}} \approx 4\ \text{kcal g}^{-1} = 4\ \text{Cal g}^{-1}$$

5.0 g then provides about 20 Cal and the manufacturer's claim <u>is correct</u>.

Box 7.2, Exercise 1

In order to calculate the fractional saturation at the requested pressure we need to know the value of K in the Hill equation. K can be determined approximately from the figure shown in the Box. For $S = 0.5, p = p_{50}$ where p_{50} is the pressure of O_2 at 50% saturation.

$$\log\left(\frac{0.5}{1-0.5}\right) = \nu \log p_{50} - \nu \log K$$

Therefore $K = p_{50}$.
From the graph we estimate that for hemoglobin, $p_{50} = K = 26$ Torr; for myoglobin, $p_{50} = K = 5$ Torr. Solving the Hill equation for s we obtain

$$\frac{s}{1-s} = \left(\frac{p_{O_2}}{K}\right)^{\nu}$$

and

$$s = \frac{1}{\left(\frac{K}{p_{O_2}}\right)^{\nu} + 1}$$

Then for Hb we construct the following table.

p_{O_2}/Torr	5	10	20	30	60
s	9.8×10^{-3}	0.064	0.32	0.60	0.91

For Mb

p_{O_2}/Torr	5	10	20	30	60
s	0.50	0.66	0.80	0.86	0.92

The values of S match well the values read from the graph.

Box 9.1, Exercise 1

The reaction referred to is presumably
$ATP(aq) + H_2O(l) \rightarrow ADP(aq) + P_i(aq) + H^+(aq)$

$\Delta_r G^{\oplus}$ for this reaction is -31.3 kJ mol^{-1} (see Box 9.2). $\Delta_r G^{\oplus}$ will change slightly from the biological standard state of pH = 7 to pH = 7.1. The change can be calculated from

$$\begin{aligned}\Delta(\Delta_r G^{\oplus}) &= -2.303\, RT\, \Delta\text{pH} \\ &= -2.303 \times 8.314 \text{ J K}^{-1}\text{ mol}^{-1} \times 310 \text{ K} \times 0.1 \\ &= -6 \times 10^2 \text{ J mol}^{-1} = -0.6 \text{ kJ mol}^{-1}\end{aligned}$$

Thus $\Delta_r G$ has changed from $-31.3\ \text{kJ mol}^{-1}$ to $-31.9\ \text{kJ mol}^{-1}$. This is insignificant.

If the pH in the intermembrane space increases by 0.1, then ΔpH changes from -1.4 (see Box 9.2) to -1.3. ΔG_m due to the concentration gradient is now

$$\begin{aligned}\Delta G_m &= -2.303\ RT\,\Delta\text{pH} = -2.303\ RT \times (-1.3)\\ &= -2.303 \times 8.314\ \text{J K}^{-1}\ \text{mol}^{-1} \times 310\ \text{K} \times (-1.3)\\ &= +7.7\ \text{kJ mol}^{-1}\end{aligned}$$

The overall ΔG_m including the contribution from the potential difference, which is assumed to remain the same at 0.14 V, is then

$$\begin{aligned}\Delta G_m &= +7.7\ \text{kJ mol}^{-1} + 13.5\ \text{kJ mol}^{-1}\\ &= +21.2\ \text{kJ mol}^{-1}\end{aligned}$$

For the transport of 2 mol H^+, $\Delta G = +41.2$ kJ. Therefore the amount of ATP that is required is

$$\frac{41.2\ \text{kJ}}{31.9\ \text{kJ mol}^{-1}} = \underline{1.29\ \text{mol}}$$

Box 9.2, Exercise 1

−70 mV		
$[Na^+] = 10\ \text{mmol L}^{-1}$	membrane	$[Na^+] = 140\ \text{mmol L}^{-1}$
$[K^+] = 100\ \text{mmol L}^{-1}$		$[K^+] = 5\ \text{mmol L}^{-1}$

$$\begin{aligned}\Delta G(Na^+) &= G(Na^+, \text{out}) - G(Na^+, \text{in}) = RT\ln Q + F\Delta\phi\\ &= RT\ln\frac{[Na^+]_{out}}{[Na^+]_{in}} + F\Delta\phi\\ &= 8.314\ \text{J K}^{-1}\ \text{mol}^{-1} \times 310\ \text{K} \times \ln\left(\frac{140}{10}\right) + 9.6485 \times 10^4\ \text{C mol}^{-1} \times 0.070\ \text{V}\\ &= +1.36 \times 10^4\ \text{J mol}^{-1} = \underline{+13.6\ \text{kJ mol}^{-1}}\end{aligned}$$

$$\begin{aligned}\Delta G(K^+) &= G(K^+, \text{in}) - G(K^+, \text{out}) = RT\ln Q + F\Delta\phi\\ &= RT\ln\frac{[K^+]_{out}}{[K^+]_{in}} + F\Delta\phi\\ &= 8.314\ \text{J K}^{-1}\ \text{mol}^{-1} \times 310\ \text{K} \times \ln\left(\frac{5}{100}\right) + 9.6485 \times 10^4\ \text{C mol}^{-1} \times (-0.070\ \text{V})\\ &= \underline{+1.0\ \text{kJ mol}^{-1}}\end{aligned}$$

Box 9.3, Exercise 1

(a) $\Delta_r G^{\ominus} = -\nu F E^{\ominus} \qquad \nu = 2$

$$\Delta_r G^{\ominus} = -2 \times 9.6485 \times 10^4 \text{ C mol}^{-1} \times 0.562 \text{ V}$$
$$= \underline{1.085 \times 10^5 \text{ J mol}^{-1}}$$

(b) $K = e^{-\Delta_r G^{\ominus}/RT} = e^{-(1.085 \times 10^5/8.3145 \times 298)}$

$$= \underline{9.60 \times 10^{-20}}$$

Box 10.1, Exercise 1

[CO] changes little during the course of the reaction for the concentration given, so [Mb] follows pseudofirst order kinetics.

$[\text{Mb}] = [\text{Mb}]_0 e^{-k't}$

k' is the rate constant for the pseudofirst-order process. The rate constant, k, which is given, is for the second order process. That is

rate $= k[\text{CO}][\text{Mb}] = k'[\text{Mb}]$

$$k' = k[\text{CO}]$$
$$= 5.8 \times 10^5 \text{ L mol}^{-1} \text{ s}^{-1} \times 0.400 \text{ mol L}^{-1}$$
$$= 2.3 \times 10^5 \text{ s}^{-1}$$

A curve of [Mb] against time using this value of k' is shown in the figure below.

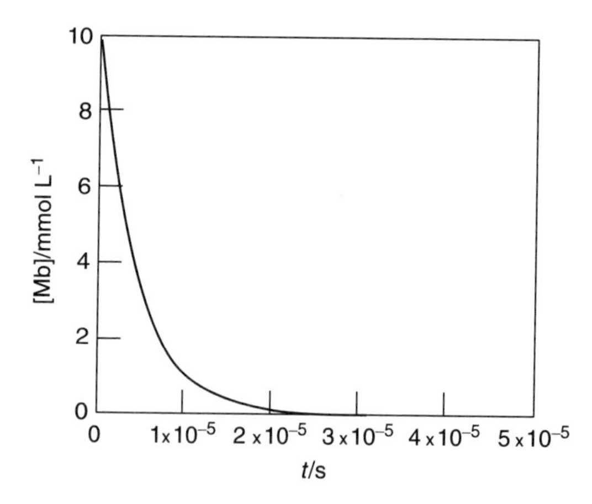

Box 11.1, Exercise 1

Assume the reaction is first order, then

$$\frac{N}{N_0} = e^{-kt} \qquad k = \frac{\ln 2}{t_{1/2}}$$

$N = 10^9 - 1$

$N_0 = 10^9$

$$\ln\left(\frac{N}{N_0}\right) = -kt$$

$$t = \frac{\ln\left(\frac{N}{N_0}\right)}{-k} = \frac{\ln\left(\frac{10^9 - 1}{10^9}\right)}{-k} = \frac{1 \times 10^{-9}}{k}$$

$$t = \frac{1 \times 10^{-9}\, t_{1/2}}{\ln 2} = 1.44 \times 10^{-9}\, t_{1/2}$$

$$= 1.87 \times 10^{-4}\ \text{y} = \underline{1.64\ \text{min}}$$

Not very long!

Box 11.2, Exercise 1

$$\tilde{\nu} = \frac{1}{\lambda} = \frac{1}{2.60 \times 10^{-7}\ \text{m}} = 3.85 \times 10^6\ \text{m}^{-1}$$

$$= 3.85 \times 10^4\ \text{cm}^{-1}$$

$$E = N_A \times 3.85 \times 10^4\ \text{cm}^{-1} \times 1.9864 \times 10^{-23}\ \text{J cm}$$

$$\times \frac{1\ \text{kJ}}{10^3\ \text{J}} \times \frac{1\ \text{kcal}}{4.184\ \text{kJ}} = \underline{110\ \text{kcal mol}^{-1}}$$

Box 12.1, Exercise 1

(a) First evaluate the factor

$$\left(1 + \frac{eV}{2m_e c^2}\right) = \left(1 + \frac{1.602 \times 10^{-19}\ \text{C} \times 5.0 \times 10^4\ \text{V}}{2 \times 9.11 \times 10^{-31}\ \text{kg} \times (2.998 \times 10^8)^2}\right)$$

$$= 1.049$$

$$\text{Then, } \lambda = \frac{6.626 \times 10^{-34}\ \text{J S}}{(2 \times 9.11 \times 10^{-31}\ \text{kg} \times 1.602 \times 10^{-19}\ \text{C} \times 5.0 \times 10^4\ \text{V} \times 1.049)^{1/2}}$$

$$= 5.36 \times 10^{-12}\ \text{m} = \underline{5.36\ \text{pm}}$$

(b) It makes about 2.5% difference; see part (a).

Box 13.1, Exercise 1

(a) Zn^{2+} $1s^22s^22p^63s^23p^63d^{10}$

(b) Zn^{2+} has no empty orbitals at $n = 3$. So electrons in O_2 have no place to go at this level in order to form a bond with Fe^{2+} of the haemgroup. They could occupy empty orbitals at $n = 4$, but this is energetically unfavorable.

Box 14.1, Exercise 1

In a vacuum $\varepsilon = \varepsilon_0$. Therefore

$$V = \frac{-(1.602 \times 10^{-19}\ \text{C})^2}{4\pi \times 8.854 \times 10^{-12}\ \text{J}^{-1}\ \text{C}^2\ \text{m}^{-1} \times 5.292 \times 10^{-11}\ \text{m}}$$

$$= \underline{-4.36 \times 10^{-14}\ \text{J}}$$

(a) We need to know the partial charges on H and O in the hydrogen bonded group and the N—H and O=C bond distances. The partial charges can be found in Table 16.1 and are: for H(–N), +0.18 e; for O, –0.38 e. The bond distances in a peptide link are: for N—H, 102 pm; for O=C, 124 pm. If we assume a linear arrangement of the bonded groups, this results in a H ...O distance of 290 pm –102 pm = 188 pm. Therefore, for the hydrogen bond in a membrane

$$V = \frac{-0.38 \times 0.18 \times (1.602 \times 10^{-19}\ \text{C})^2}{4\pi \times 2 \times 8.854 \times 10^{-12}\ \text{J}^{-1}\ \text{C}^2\ \text{m}^{-1} \times 1.88 \times 10^{-10}\ \text{m}}$$

$$= \underline{-4.2 \times 10^{-20}\ \text{J}}$$

(b) In water, $\varepsilon = 80 \times \varepsilon_0$

$$V = \underline{-1.0 \times 10^{-21}\ \text{J}}$$

Box 16.1, Exercise 1

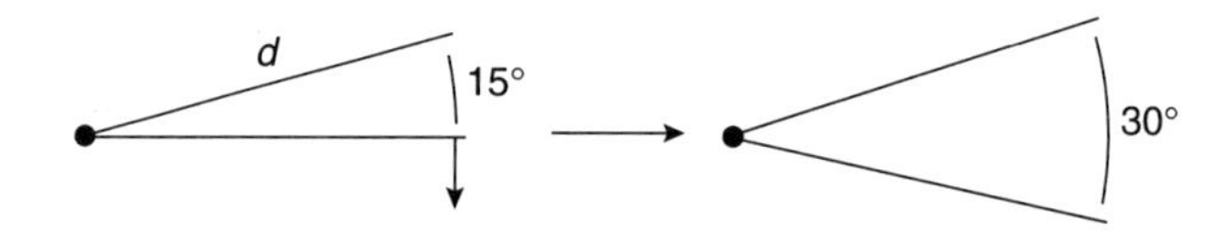

Consider the situation described in the sketch above. The application of a force through the distance indicated by the downward arrow causes the bending angle to increase from 15° to 30° with the expenditure of 8.5 kJ mol^{-1} of work. The distance moved is

$$x = \frac{\Delta\theta}{2\pi} \quad d = \frac{15°}{360°}\, d$$

The "bending" force constant is then given by

$$F = -k'x = -k'\left(\frac{d}{2\pi}\right)\Delta\theta = -k\Delta\theta$$

because d and 2π are constants. The energy expended is then given by

$$V = \frac{1}{2}k'x^2 = \frac{1}{2}(2\pi k/d)\Delta\theta = 8.5 \text{ kJ mol}^{-1}$$

but without knowledge of d, k cannot be evaluated. Estimates of d could be made if the structure of the molecule were shown. Short of that, we can evaluate an effective force constant for bending (in units of J deg^{-2}) from

$$V = \frac{1}{2}k_{\text{bend}}(\theta - \theta_e)^2$$

$$k_{\text{bend}} = \frac{2V}{(\theta - \theta_e)^2} = \frac{2 \times 8.5 \times 10^3 \text{ J mol}^{-1}}{6.02 \times 10^{23} \text{ mol}^{-1} \times (15 \text{ deg})^2}$$

$$= \underline{1.3 \times 10^{-22} \text{ J deg}^{2}}$$

Box 16.2, Exercise 1

In both cases, the diffusion is two-dimensional. Therefore,

$l = (4Dt)^{1/2}$

and

$$t = \frac{l^2}{4D}$$

In a cell plasma membrane

$$t = \frac{(1.0 \times 10^{-8} \text{ m})^2}{4 \times 1.0 \times 10^{-8} \text{ cm}^2 \text{ s}^{-1} \times 1 \text{ m}^2/10^4 \text{ cm}^2}$$

$$= \underline{2.5 \times 10^{-5} \text{ s}}$$

In a lipid bilayer

$t = \underline{2.5 \times 10^{-6} \text{ s}}$

Box 18.1, Exercise 1

Percentage transmitted to the retina is

$$70\% - 0.25 \times 70\% - 0.09 \times 0.75 \times 70\% - 0.43 \times (1 - 0.25 - 0.09 \times 0.75) \times 70\%$$

$$= 70\% - (0.25 + 0.0675 + 0.390) \times 70\%$$

$$= 20.5\%$$

Number of photons focused on the retina in 0.1 s is

$0.205 \times 40\ \text{mm}^2 \times 0.1\ \text{s} \times 4 \times 10^3\ \text{mm}^{-2}\ \text{s}^{-1}$

$= \underline{3 \times 10^3}$

More than what one might have guessed.

Box 19.1, Exercise 1

We use $\nu = \dfrac{\gamma_N \mathcal{B}_{loc}}{2\pi} = \dfrac{\gamma_N}{2\pi}(1 - \sigma)\mathcal{B}$ [19.6]

where $\mathcal{B}$ is the applied field.

Because shielding constants are quite small (a few parts per million) compared to 1, we may write for the purposes of this calculation

$$\nu = \frac{\gamma_N \mathcal{B}}{2\pi}$$

$$\nu_L - \nu_R = 100\ \text{Hz} = \frac{\gamma_N}{2\pi}(\mathcal{B}_L - \mathcal{B}_R)$$

$$\mathcal{B}_L - \mathcal{B}_R = \frac{2\pi \times 100\ \text{s}^{-1}}{\gamma_N}$$

$$= \frac{2\pi \times 100\ \text{s}^{-1}}{26.752 \times 10^7\ \text{T}^{-1}\ \text{s}^{-1}} - 2.35 \times 10^{-6}\ \text{T}$$

$$= 2.35\ \mu\text{T}$$

The field gradient required is then

$$\frac{2.35\ \mu\text{T}}{0.08\ \text{m}} = \underline{29\ \mu\text{T m}^{-1}}$$

Note that knowledge of the spectrometer frequency, applied field, and the numerical value of the chemical shift (because constant) is not required.

Box 20.1, Exercise 1

(1) glycine

$$\alpha = \frac{1}{2}\left\{1 + \frac{(0.62 - 1) + 2 \times 1.0 \times 10^{-5}}{[(0.62 - 1)^2 + 4 \times 0.62 \times 1.0 \times 10^{-5}]^{1/2}}\right\}$$

$$= \underline{6.9 \times 10^{-5}}$$

(2) L-leucine

$$\alpha = \frac{1}{2}\left\{1 + \frac{(1.14 - 1) + 2 \times 33 \times 10^{-4}}{[(1.14 - 1)^2 + 4 \times 1.14 \times 33 \times 10^{-4}]^{1/2}}\right\}$$

$$= \underline{0.89}$$